Advances in Anatomy Embryology and Cell Biology

Vol. 169

Springer
Berlin
Heidelberg
New York
Hong Kong
London
Milan
Paris
Tokyo

P.H. Francis-West L. Robson
D.J.R. Evans

Craniofacial Development
The Tissue and Molecular Interactions That Control Development of the Head

With 17 Figures and 5 Tables

Springer

Dr. P.H. Francis-West
Department of Craniofacial Development
King's College
London, SE1 9RT, UK
e-mail: pfrancis@hgmp.mrc.ac.uk

Dr. L. Robson
Department of Neuroscience
Bart's and The London
Queen Mary's School of Medicine and Dentistry
London, E1 4NS, UK
e-mail: L.G.Robson@qmul.ac.uk

Dr. Darell J.R. Evans
School of Biosciences
Cardiff University
Cardiff, CF10 3US, UK
e-mail: EvansDJR@Cardiff.ac.uk

ISSN 0301-5556
ISBN 3-540-00363-0 Springer-Verlag Berlin Heidelberg NewYork

Library of Congress-Cataloging-in-Publication-Data
Francis-West, Philippa H., 1964– Craniofacial development: the tissue and molecular interactions that control development of the head / Philippa H. Francis-West, Lesley Robson, Darell J.R. Evans. p.; cm. — (Advances in anatomy, embryology, and cell biology, ISSN 0301-5556; v. 169)
Includes bibliographical references and index.
ISBN 3-540-00363-0 (softcover: alk. paper)
1. Head. 2. Embryology. 3. Head—Molecular aspects. 4. Facial bones—Growth. 5. Skull—Growth. I. Robson, Lesley, 1965– II. Evans, Darrell J.R., 1969– III. Title. IV. Series

Springer-Verlag a member of BertelsmannSpringer
Science + Business Media GmbH

http://www.springer.de

Printed in Germany

Production: PRO EDIT GmbH, 69126 Heidelberg, Germany
Printed on acid-free paper 27/3150Re – 5 4 3 2 1 0

Contents

1 Introduction

Craniofacial development is a multistep and intricate process initially involving a number of inductive interactions that control neural and neural crest development, which are followed by a series of epithelial-mesenchymal interactions that control outgrowth, patterning, and skeletal differentiation. Certain aspects of craniofacial development are unique developmental processes in higher vertebrates. First, in higher vertebrates the cranial neural crest, in contrast to the trunk neural crest, gives rise to the skeletal structures. These skeletal elements include those comprising membrane bone and secondary cartilage, which with the exception of the clavicle are tissue types found exclusively in the head in higher vertebrates. Second, with the exception of the tongue, the origin of the musculature is distinct from other regions of the body. The body and tongue muscles are formed from the segmented epithelial somites whilst the head musculature is formed from unsegmented paraxial and prechordal mesoderm. Furthermore, the signalling cascades that control myogenic differentiation appear to be distinct as determined by gene expression and the response of myogenic cells to growth factors. Finally, the neurogenic placodes, which give rise to the sensory organs and some cranial ganglia, are only found in the head. Over recent years, there have been significant advances in our knowledge of the molecular processes that control craniofacial development in a number of animal models. This has given insight into the genes that control many aspects of head development from the initial induction of the head to the final stages of differentiation. In addition, rapid progress has been made in the identification of the gene mutations/deletions responsible for human syndromes such as DiGeorge and Riegers syndromes. This has put the field of craniofacial development at a position where knowledge of signalling interactions that control the development of the head will exponentially increase, giving insight into not only how the intricate structures of the head are patterned but also its evolution. As mentioned above, the head is a unique structure, with a number of distinct features. Therefore, the signalling pathways are unlikely to be a simple repeat of those found in other regions of the body such as the limb and the somites. This review will discuss current knowledge of the molecular and tissue interactions that control craniofacial development, highlighting the recent advances, with reference to the human syndromes where appropriate.

2 Neural and Head Development are Intimately Linked

Embryonic and post-natal development of the head is dependent on correct neural development. First, closure of the neural plate is essential for formation of the skull whilst patterning along the midline and the rostrocaudal axis is essential for development of the face. If the neural plate does not close, there is an "open brain" phenotype known as exencephaly where the brain is exposed and may be everted. The skull bones also will not develop, a manifestation known as acrania. Failure of the anterior neural plate to divide medially can result in holoprosencephaly whilst defects in rostrocaudal patterning, such as subdivision of the hindbrain into rhombomeres, can result in abnormal development of the neural crest, which not only forms the skeletal structures of the face but gives rise to many connective tissues of the head (see Sect. 3.1). This will affect the development of the face. Signals from the neural tube are also essential for subsequent proliferation and survival of the neural crest cells (see Sects. 4.1, 4.4). Finally, post-natally, development of the brain is linked to the growth of the skull (see Sect. 6.1). As the brain grows the skull bones are separated, which then continue to grow at sutural edges to fill the gap, a process known as displacement. This growth of the brain, together with development of the eye and paranasal air sinuses, will also contribute to post-natal growth of the face, which at birth is relatively underdeveloped.

2.1 Exencephaly

Exencephaly in mice and anencephaly in humans are the failure of the neural tube to close resulting in the absence of skin and bone over the brain (acrania). In some cases the germinal or inner layer of the brain faces outwards, known as cranioschisis. Anencephaly occurs at an approximate frequency of 1/1,000 births worldwide. However, in mice and humans, females are more susceptible (Lemire 1988; MacDonald et al. 1989). The causes are multifactorial, with approximately 80% being random events due to environmental and/or contributory genetic factors. In the latter case, the genes responsible for anencephaly or which predispose individuals are starting to be identified. For example, Knobloch syndrome, which is in part characterised by occipital exencephaly, is due to truncation of $\alpha 1$ (XVIII) collagen, a basement membrane component (Sertié et al. 2000). The defects can range from delayed closure as seen in the *Bmp5/7* double mouse mutants to partial closure such as in the *open brain* (*opb*) mouse mutants, which have a mid-forebrain closure defect, to failure of all of the cranial neuroepithelium to close as occurs in the *AP-2* mutants (Günther et al. 1994; Schlore et al. 1996; Zhang et al. 1996; Solloway and Robertson 1999). The mouse mutants which have an exencephalic phenotype are listed in Table 1 (also see review by Juriloff and Harris 2000). Administration of folic acid has been shown to significantly decrease the frequency of neural tube abnormalities and can even rescue the exencephalic defects seen in the *Cart1* mutant (Zhao et al. 1996; reviewed by van der Put et al. 2001). In general the mechanisms of folate rescue are unknown. However, in some cases it can compensate for mutations in *methylenetetrahydrofolate reductase*, an enzyme involved in folate metabolism (reviewed by van der Put et al. 2001).

Neural tube closure is dependent on cell shape changes and migration together with the correct levels of cell proliferation and survival. During normal development, the neural plate first pivots along the midline at a region called the medial hinge point (MHP) (Fig. 1). Defects in this process occur in the *loop-tail* mouse mutant, which has an abnormally wide flat midline (Greene et al. 1998). Subsequently, at the presumptive brain level (and the caudal neuropore), the neural cells adjacent to the ectoderm become elongated and wedge-shaped being wider at the basal surface, such that the neural groove bends medially to form the dorsal lateral hinge points (DLHPs) (Fig. 1; reviewed by Smith and Schoenwolf 1997). In addition, intercalation of neural cells, which simultaneously narrows the neural plate along the medial-lateral axis whilst elongating it along the anterior-posterior axis, may contribute to neural plate closure by moving the neural folds into closer apposition. The lateral edges of the neural fold meet and adhere through cell surface glyconjugates and cellular protrusions. The neural cells fuse, with some undergoing apoptosis, to form a seamless tube, the neural tube (Schlüter 1973). Similarly, the remaining overlying ectoderm edges meet and fuse and this layer will ultimately form the skin (Fig. 1).

The positions and numbers of closure points in the neural tube, together with the rates of neural tube closure vary among species. In mice the cranial neural fold meets at three points: first, at the hindbrain/spinal cord boundary, then usually at the fore/midbrain junction, and subsequently at the most rostral region of the forebrain (Fig. 2; Geelen and Langman 1977; Sakai 1989; Juriloff et al. 1991; van Allen et al. 1993; Golden and Chernoff 1995). The position of the closure 2 site has been shown to be variable and can be rostral or caudal to the forebrain-midbrain boundary, as in the SWV/Bc or DBA/2 mice strains, respectively (Fig. 2). The mouse strain SELH/Bc appears to lack closure site 2 totally, although it is possible that it has a very rostral closure 2 point. These differences have been correlated with the different susceptibilities of various mouse strains to exencephaly (Juriloff et al. 1991; Fleming and Copp

Fig. 1. Diagram of the formation of the neural tube. The notochord stimulates the overlying ectoderm to form a thickened layer of cells called the neural plate. In response to further signalling from the notochord and also from the flanking ectoderm, the neural plate invaginates using the median hinge plate cells (*MHPs*) as a focus. Two dorsolateral hinge points (*DLHPs*) facilitate further folding of the neural plate, a process promoted by extrinsic forces generated by the ectoderm. Neural tube formation is completed as the edges of the neural plate (folds) appose one another and eventually fuse. Some of the factors that control cranial neural tube closure are marked, such as Abl/Arg, which are expressed at the apical surface of the closing neural plate, laminin, which is required in the ectoderm, and twist, which is needed in the cranial mesenchyme. See Sect. 2.1 for further details

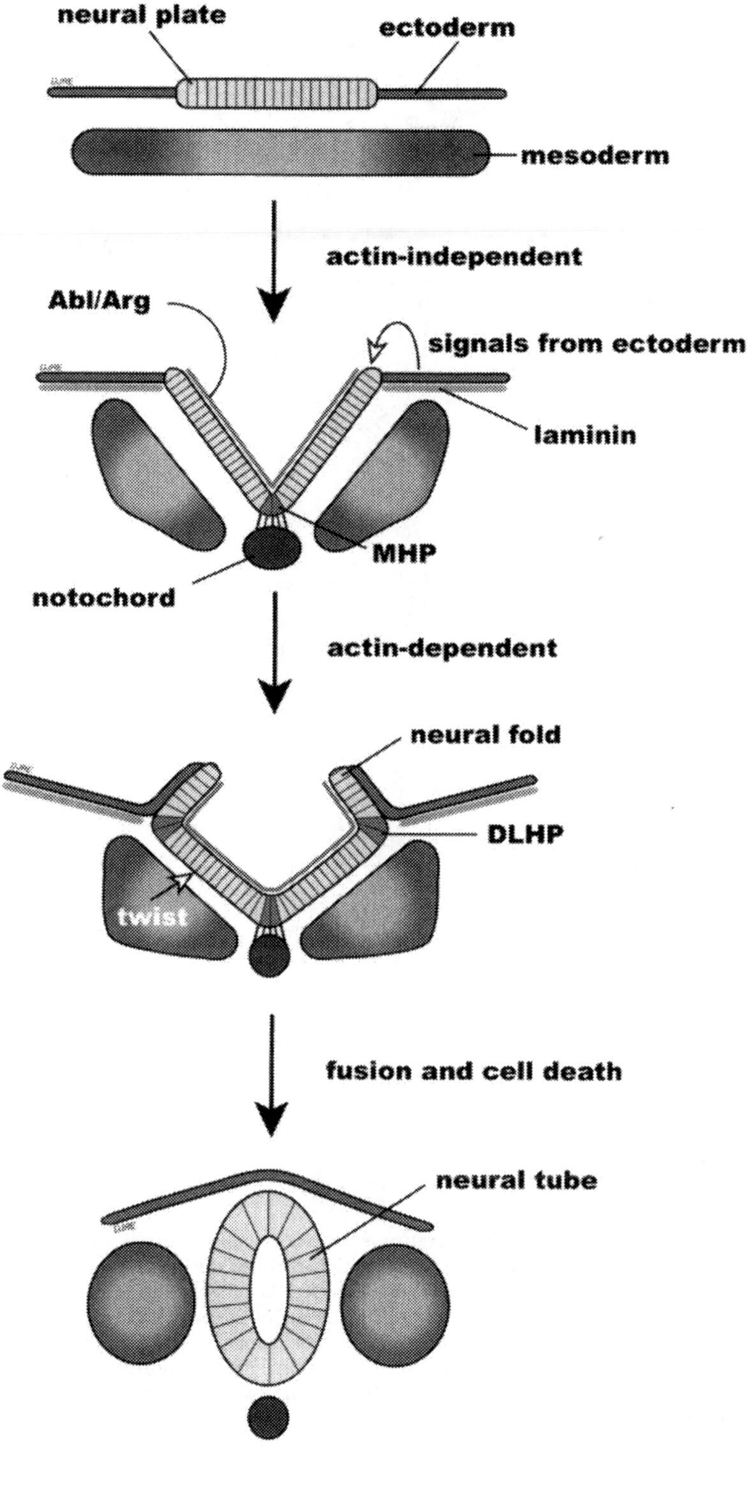
neural plate
ectoderm
mesoderm
actin-independent
Abl/Arg
signals from ectoderm
laminin
MHP
notochord
actin-dependent
neural fold
DLHP
twist
fusion and cell death
neural tube

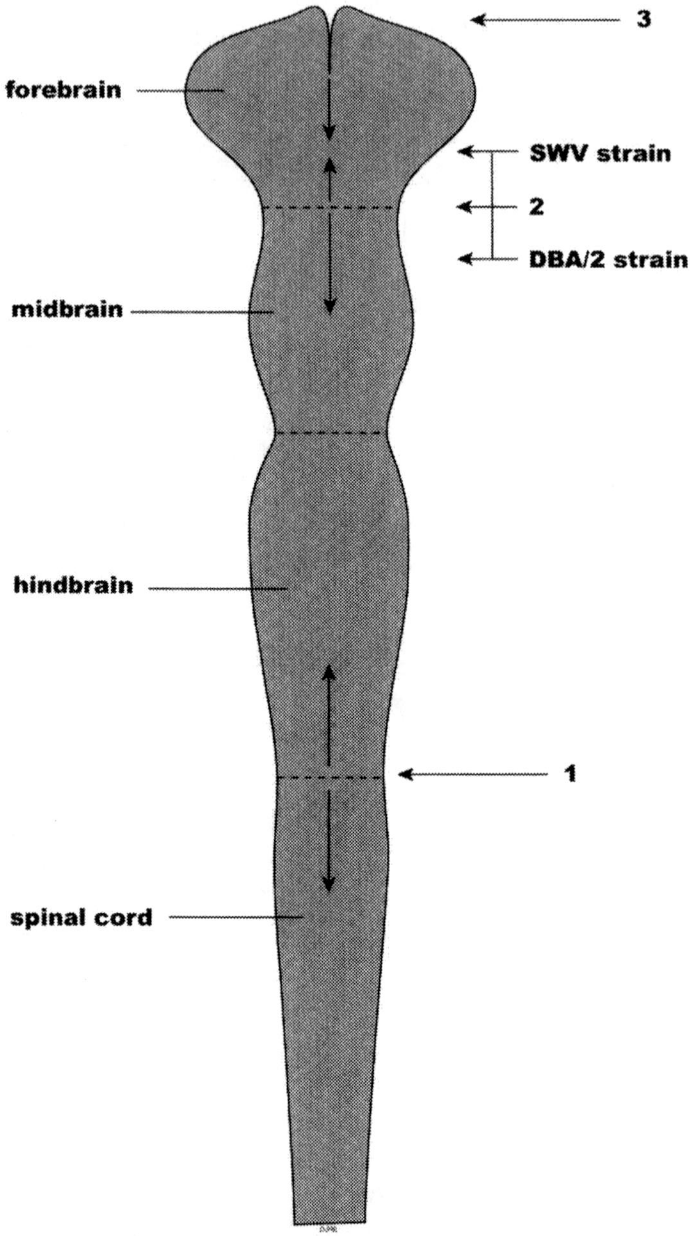

Fig. 2. Diagram of sites of cranial neural tube closure in mice. Closure site 1 is initiated at the hindbrain-cervical boundary and spreads bidirectionally from this point. Closure site 2 starts at the boundary of the forebrain and the midbrain and spreads caudally to closure site 1 and rostrally towards closure site 3. Closure site 3 is initiated at the end of the neural plate and is completed at the anterior neuropore. Closure site 2 can be initiated either more caudally as is the case in the DBA/2 strain of mice or more rostrally as in SWV mice. The variations in the position of the closure 2 site are correlated with the different susceptibilities of mouse strains to exencephaly, as strains with a more rostral closure 2 site are less susceptible to exencephaly

2000). For example, the SWV/Bc strain is less susceptible to exencephalic defects than the SELH/Bc mice, in which there is 17% failure of neural tube closure. These differences are also highlighted by the susceptibility of different strains of mice to cranial neural tube defects due to loss of a gene. Following loss of the *jumonji* gene, exencephaly occurs on a C3H/HeJ or mixed BALB/cA/129/Ola background but not on BALBc/cA, C57BL/6J, and DBA/2J backgrounds (Takeuchi et al. 1995, 1999). In humans, closure points 1 and 3 have been identified but the more variable closure 2 point seen in mice has not been observed. In the chick there is a different sequence of events, with the neural folds first coming into apposition at the hindbrain-midbrain junction followed by closure at the hindbrain-spinal cord boundary (Anderson and Meier 1981; Jaskoll et al. 1991).

The initial cell shape changes and furrowing of the neural plate are intrinsic to the neural plate whereas for development of the DLHPs, extrinsic signals are required (Schoenwolf and Smith 1990). These include changes in morphogenetic movements such as the development of the cranial flexure and rotation of the body axis. In particular the ectoderm, adjacent to the neural fold, appears to be a prime and active force during formation of the DLHPs. If this region of ectoderm is removed, the neural folds cannot close (Hackett et al. 1997; Schoenwolf and Smith 1990). In addition, signals from the basal lamina, the extracellular matrix coating of cells, are essential as shown by the exencephalic phenotype following loss of function of *laminin* $\alpha 5$, a component of basal laminae (Fig. 1; Miner et al. 1998). In these mutants the basal lamina underlying the ectoderm adjacent to the neural folds is thin and non-uniform. The defect may reflect a signalling function (e.g., through integrins) and/or a mechanical role of laminin.

Studies using cytochalasin D, which destablises actin filaments, have shown that the cranial DLHPs, but not the MHPs and the caudal DLHPs, require actin for their formation (Morriss-Kay and Tucket 1985; Schoenwolf et al. 1988; Ybot-Gonzalez and Copp 1999). Actin function is thought to be important for the changes in the shape of cells, which become taller and wedge-shaped (Fig. 1). The role of actin filaments in this process has been more recently confirmed by the generation of mouse mutants lacking proteins associated with the actin cytoskeleton, such as Mena, profilin 1, Abl, Arg, p190RhoGAP, MARCKS (myristoylated alanine-rich C kinase), and MacMarcks (also known as MRP and F52). All these proteins link signal transduction events with changes in actin polymerisation, and all the homozygous mouse mutants have exencephaly (Fig. 1, Table 1; Stumpo et al. 1995; Chen et al. 1996; Wu et al. 1996; Koleske et al. 1998; Lanier et al. 1999; Brouns et al. 2000). In *Abl/Arg* double mutants, in addition to disorganisation of the actin cytoskeleton, there is increased apoptosis in all tissues, including the developing brain, at E10.5–11, which may also be a contributory factor to the exencephalic phenotype (Koleske et al. 1998).

Again, emphasising the necessity of actin function, Shroom, which is found in adherens junctions and binds actin stress fibres, is also essential for cranial neural tube closure. *Shroom* is expressed in the cranial neuroepithelium at E8.75, and the defects in *Shroom* mutants are apparent at E9.5. In this mutant, neural cell proliferation and survival are normal but, as also occurs in the *MacMARCKs* mutant, the neural folds grow outwards, apparently failing to form the DLHPs (Chen et al. 1996; Wu et al. 1996; Hildebrand and Soriano 1999).

The correct balance of cell proliferation and survival is also crucial for neural tube morphogenesis. Cell death is predominantly found in the neural folds and is critical

Table 1. The mouse mutants that have exencephaly[a]

Gene	Expression	Phenotype	Mechanism	Reference
Transcription factors				
Ap2	Neural folds and adjacent ectoderm	Exencephaly Face defects	Increase in apoptosis in hindbrain Ectopic expression of N-CAM	Schlore et al. 1998 Zhang et al. 1996
Bent tail mutant (deletion in Zic3)	Neuroepithelium	Exencephaly (Hindbrain) Face defects	Decrease in proliferation?	Klootwijk et al. 2000
Cart1	Cranial mesenchyme Skeleton	Exencephaly (Midbrain) Face defects	Increase in apoptosis in mesenchyme underlying forebrain at E8.75 Transient decrease in number of mesenchyme cells underlying forebrain at E8.75 Neural folds elevate but do not bend towards each other	Zhao et al. 1996
Dlx5	Neuroepithelium Surface ectoderm	Exencephaly Face defects	Not known	Acampora et al. 1999 Depew et al. 1999
Gli3 mutant	Facial mesenchyme Skeleton	Exencephaly Face defects	Not known	Hui and Joyner 1993 Mo et al. 1997
Hic1	Ubiquitous	Exencephaly Face defects	Not known	Carter et al. 2000
Hes1	Neuroepithelium Cranial mesenchyme	Exencephaly Face defects	Neural folds elevate but do not fuse	Ishibashi et al. 1995
δEF1	Headfold and cephalic mesenchyme Skeleton	Exencephaly Face defects	Not known	Funahashi et al. 1993 Takagi et al. 1998
jumonji mutant (transcription factor?)	Rostral neural plate	Exencephaly	Abnormal medial-lateral groove at midbrain-hindbrain boundary	Takeuchi et al. 1999
Ski	Neuroepithelium Migrating neural crest	Exencephaly (Midbrain–forebrain boundary) Face defects	Increase in apoptosis in neuroepithelium and cranial mesenchyme Levels of proliferation are unchanged Cells sparser in mesenchyme	Berk et al. 1997

Table 1. *Continued.*

Gene	Expression	Phenotype	Mechanism	Reference
Twist	Cranial mesenchyme	Exencephaly (Anterior hindbrain to forebrain) Face defects	Mesenchymal defects – cells sparser Cell adhesion effect? Neural folds elevate and bend towards each other but do not fuse Neuroepithelium degenerates	Chen and Behringer, 1995
Yin Yang (Zinc finger transcription factor)	Cranial region at E7.5 Ubiquitous	Exencephaly (Heterozygotes)	Not known	Donohoe et al. 1999
p53		Exencephaly (Midbrain)	Not known	Sah et al. 1995
Matrix molecules				
Laminin $\alpha 5$	Neural Folds Surface ectoderm	Exencephaly	Abnormal basal lamina underlying ectodermal cells next to neural tube	Miner et al. 1998
Perlecan		Exencephaly Face defects	Not known	Arikawa-Hirasawa et al. 1999
Intracellular proteins				
Abl/Arg	Neuroepithelium (apical surface)	Exencephaly	Involved in apical constriction? Ectopic actin at basal surface of neural tube Levels of proliferation are unchanged Increase in apoptosis Neural tube starts to close but buckles inwards	Koleske et al. 1998
IKK1/2		Exencephaly (Hindbrain)	Increase in apoptosis in hindbrain at E9.5 Neural folds do not elevate at midline or bend towards each other	Kuan et al. 1999 Li et al. 1999
Apaf-1	Neuroepithelium	Exencephaly (Rostral) Face defects	Decrease in apoptosis Increase in levels of proliferation Brain hyperplasia – neuroepithelium thicker at E9.5	Cecconi et al. 1998 Honarpour et al. 2000 Yoshida et al. 1998

Table 1. *Continued.*

Gene	Expression	Phenotype	Mechanism	Reference
Bcl10		Exencephaly (Hindbrain)	Increase in apoptosis in hindbrain at E10.5 Neural folds do not elevate or bend towards each other	Ruland et al. 2001
Caspase9		Exencephaly (Hindbrain)	Decrease in apoptosis Brain hyperplasia	Hakem et al. 1998 Kuida et al. 1996
JNK1/2	JNK1/2 activity detected in neuroepithelium	Exencephaly (Hindbrain)	Decrease in apoptosis in hindbrain Increase in apoptosis in forebrain	Kuan et al. 1999
Numb	Neuroepithelium	Exencephaly (All cranial region)		Zhong et al. 2000
Nup50	Ubiquitous but higher in neural tube	Exencephaly	Ectopic p27Kip1 expression in neuroepithelium	Smitherman et al. 2000
MacMARCKs	Neural folds MHP Cranial mesenchyme	Exencephaly (Hindbrain to forebrain) Hypotelorism	DLHPs do not form Neural folds splay outwards	Chen et al. 1996 Wu et al. 1996
MARCKs		Exencephaly Microcephaly	Not known	Stumpo et al. 1995
Mena/profilin	Neuroepithelium	Exencephaly	DLHP formation? Interacts with Abl/Arg	Lanier et al. 1999
Shroom (novel PDZ domain protein)	Neuroepithelium	Exencephaly (All cranial region) Face defects	Levels of apoptosis and proliferation unchanged Binds to and changes localisation of F-actin Neural folds elevate but do not bend towards each other	Hildebrand and Soriano 1999
Telomerase RNA	Neuroepithelium	Exencephaly (Midbrain to forebrain)	Increase in apoptosis in neuroepithelium at E8.5 Levels of proliferation are unchanged	Herrera et al. 1999
Treacle	Neuroepithelium	Exencephaly Face defects	Increase in apoptosis in neuroepithelium	Dixon et al. 2000

Table 1. *Continued.*

Gene	Expression	Phenotype	Mechanism	Reference
p190RhoGAP	Neuroepithelium	Exencephaly (Hindbrain to forebrain)	Levels of apoptosis unchanged Excess F-actin accumulates in MHP region MHP region is wider Cell shape and integrin effects?	Brouns et al. 2000
p300	Ubiquitous	Exencephaly (Hindbrain to forebrain)	Decrease in cell proliferation	Yao et al. 1998
loop-tail (lp) mutant (ltap, involved in fz/dsh pathway)	Neuroepithelium	Exencephaly (Rostral to midbrain)	MHP region is wider Neural folds do not elevate Does not initiate closure at cervical/hindbrain boundary	Greene et al. 1998 Kibar et al. 2001
Unidentified				
118 ENU-induced mutant (same as open brain mutant)			Neural tube dorso-ventral patterning defect	Günther et al. 1994 Kasarskis et al. 1998
25 ENU-induced mutant		Exencephaly	Not known	Kasarskis et al. 1998
105 ENU-induced mutant		Exencephaly	Mesenchymal defect Cells underlying neural plate are widely spaced, small, and spindle-shaped Does not complement with Twist	Kasarskis et al. 1998

[a]The mutants are divided under the subheadings "Transcription factors," "Matrix molecules," and "Intracellular molecules," in alphabetical order within each section. The expression domain reported is just restricted to the craniofacial tissues, in particular those involved in neural tube closure, i.e. the neuroepithelium, ectoderm, and mesoderm. The stage of the development in which the defect occurs and/or the mechanism is listed under the heading "Mechanism." Other craniofacial defects are simply referred to as face defects, and more information about these is found in Table 3.

for neural fold fusion (Geelen and Langman 1977; Hoving et al. 1990). The latter has been demonstrated by the addition of caspase inhibitors, which prevent apoptosis, to the neural tube (Weil et al. 1997). Similarly, too much cell death will prevent neural tube fusion – in this case, the neural folds will be too small to appose as occurs in the *Bcl10* mutant (Ruland et al. 2001). Gene inactivation studies have shown that different regions of the brain are dependent on different signalling pathways for survival. For example, gene-inactivation of both *JNK1/2* results in increased apoptosis in the forebrain but decreased apoptosis in the hindbrain (Kuan et al. 1999). These gene-inactivation studies have also suggested that the hindbrain is particularly vulnerable to changes in cell survival/proliferation. This has been inferred from the *Bcl10, caspase-9, apoptotic protease activating factor* 1 (*Apaf1*), and *IKK1/IKK2* mutants in which the neural closure defects are restricted to the hindbrain (Table 1; Cecconi et al. 1998; Yoshida et al. 1998; Li et al. 1999; Honarpour et al. 2000). However, the mechanisms of these defects are distinct. Bcl10 and IKK1/IKK2 are components of the NFkappaβ signalling pathway, and in both of these mutants the neural folds do not elevate or bend towards each other due to increased cell death within the presumptive hindbrain (Li et al. 1999; Ruland et al. 2001). In contrast, loss of function of *Apaf1*, which induces apoptosis, or of its downstream target, the intracellular protease *caspase-9*, results in a decrease in cell death in the developing hindbrain (Kuida et al. 1998; Hakem et al. 1998; Yoshida et al. 1998; Honarpour et al. 2000). Loss of function of *caspase-3*, which is activated by caspase 9, results in a similar phenotype (Kuida et al. 1996; Woo et al. 1998). Consequently, in all three mutants, the brain overgrows. In the *Apaf1* mutant the brain forms "cauliflower-like masses" on the face whilst in the *caspase-9* mutant, mineralised tissue, presumably the remnants of the skull, is found embedded within the brain masses (Hakem et al. 1998; Kuida et al. 1998; Honarpour et al. 2000). There also appear to be changes in neuronal migration in the *Apaf1* mutant, which may contribute to the brain hyperplasia in some regions (Honarpour et al. 2000).

The transcription factors AP-2, Ski, Cart1, Treacle, and Twist also control neural cell survival, and in all of these mutants there is an increase in cell death in the neuroepithelium. AP-2 appears to control neural cell survival in both a cell-autonomous and cell non-autonomous manner (Schorle et al. 1996; Zhang et al. 1996). The defects in the *AP-2* mutants, including the changes in cell survival, may reflect changes in matrix composition and/or the consequent cell signalling. For example, N-CAM, the cell adhesion molecule, is expressed throughout the neuroepithelium in *AP-2* mutants, rather than being localised to the neural folds (Schorle et al. 1996; Zhang et al. 1996). In addition, the expression of other cell adhesion molecules may be affected. AP-2 binding sites have been identified in the promoter of *α-2 integrin*, a laminin receptor, and as discussed earlier laminin is essential for neural tube closure (Zutter et al. 1994; Miner et al. 1998). Gene-inactivation of *Ski*, which is expressed in the neural tube and migrating neural crest, results in defective mid-forebrain closure (Lyons et al. 1994; Berk et al. 1997). In addition, there are defects in the facial structures such as frontonasal clefting possibly linked to the exencephalic phenotype, a shortened flat face together with a cleft lip (Tables 1, 3; Berk et al. 1997). These defects are associated with increased cell death in both the neural tube and mesenchyme whilst proliferation is unchanged (Berk et al. 1997). As a consequence, loss of function of *Ski* affects the density of mesenchymal cells underlying the brain, which are sparser (Berk et al. 1997).

Twist and Cart1 are only required in the mesenchyme, where they act paracrinely to control neural development. In the *Twist* null mutants, neural development is initially normal and the neural folds elevate but do not fuse. This phenotype is associated with abnormalities in the underlying mesenchymal cells, where Twist is normally expressed (Fig. 1; Wolf et al. 1991; Chen and Behringer 1995; Füchtbauer 1995; Stoetzel et al. 1995; Zhao et al. 1996; Gitelman 1997). The mutant cells are rounder than wild-type cells with fewer intracellular contacts (Zhao et al. 1996). There is also local haemorrhaging, which may be a direct or secondary consequence of the loss of Twist function in the mesenchyme. In *Cart1* mutants, all the cranial neural tube is initially affected despite the fact that *Cart1* expression is specifically localised to the mesenchymal cells underlying the forebrain (Zhao et al. 1996). The neural folds elevate but formation of the DLHPs is delayed at E9. By E9.5 only the midbrain is open. In both *Twist* and *Cart1* mutants, the mesenchymal defects result in apoptosis of the neural tissue. As *Twist* and *Cart1* expression are restricted to the mesenchyme, this suggests that they regulate the expression of as yet unidentified secreted factors required for neural survival. In addition, the mesenchyme may be needed indirectly as a mechanical support for the developing neural tube. In the *Cart1* mutant, as in the *Ski* mutants, there are also defects in development of the facial structures, presumably reflecting its role in endochondral skeletal differentiation (Tables 1, 3; Berk et al. 1997). The loss of the skull skeletal elements in the *Cart1* mutants is secondary to the exencephalic phenotype.

Both excess and loss of retinoid signalling result in exencephalic phenotypes in mice (Kessel and Gruss 1991; Morriss-Kay et al. 1991; Lohnes et al. 1994; Lee et al. 1995; Ghyselinck et al. 1997; Iulianella and Lohnes 1997). For example, loss of function of the transcriptional components of the retinoic acid signalling pathway, the retinoic acid receptors (RAR or RXR), or the enzymes involved in production and degradation of retinoic acid, Raldh2 and Cyp26 respectively, all cause exencephaly (Lohnes et al. 1994; Ghyselinck et al. 1997; Wendling et al. 1999; Niederreither et al. 2000). Downstream targets of retinoid signalling include AP-2, which is essential for neural tube closure (Table 1; Schorle et al. 1996; Choi et al. 1997). Loss-of-function of p300, a transcriptional cofactor required for RAR activity, also results in exencephaly due to decreased proliferation (Yao et al. 1998). However, as p300 does not only affect or mediate RA signalling, its mechanisms of action are likely to be complex. For example, p300 also regulates the activity of other transcription factors necessary for neural tube closure, including NFkappaβ and AP-2 (Gerritsen et al. 1997).

2.2 Holoprosencephaly

Holoprosencephaly (HPE), the most common forebrain defect in humans, is characterised by the loss of midline structures and occurs in 1 in 16,000 births. In the most severe case, alobar HPE, the forebrain, or prosencephalon, is undivided along both the medial–lateral and anterior–posterior axes. Therefore, there is one single fused cerebral lobe and the telencephalon and diencephalon are not segmented into their distinct structures. There is a single eye, cyclopia, midfacial clefts, and hypoplasia of the nasal bones. At its mildest, the majority, if not all, of the brain has cleaved, the eyes are slightly closer (hypotelorism), there may be facial clefts, and/or there is a single maxillary incisor. These defects may also be associated with an impaired sense of smell (anosmia) due to abnormal development of the olfactory bulbs. Intermediate to these scenarios, semilobar HPE is similar to alobar HPE except the brain is partially divided and the interhemispheric fissure forms posteriorly (Cohen 1989a,b; Cohen and Sulik 1992). In 17% of HPE cases there are no facial abnormalities. Contributory factors to holoprosencephaly include genetic mutations, maternal diabetes (1% risk), maternal hypocholesterolemia, alcohol, and retinoic acid. A number of factors control midline development, and no one gene appears to be essential. This is emphazised by the fact that 30% of genetic carriers for mutations that can cause HPE are phenotypically normal (Cohen 1989a,b).

For normal upper face midline development, all three germ layers together with the neural crest are essential to co-ordinate the appropriate patterning. A number of animal mutant models for holoprosencephaly have been generated, which together with tissue manipulations have revealed the complexity of the signalling interactions that control midline development. These studies have revealed that midline defects can result from abnormalities in mesoderm induction, cell movement during gastrulation, a failure to maintain the anterior mesoderm, a defect in endoderm development, or defective neural development. In addition, they have identified some of the genes involved. The animal models include the mouse mutants *sonic hedgehog* (*Shh*), *megalin*, *nodal*, and *smad2*, and zebrafish mutants, *cyclops* (*cyc*), *squint* (*sqt*), *sur*, and *one-eyed pinhead* (*oep*) (see Table 1; Hatta et al. 1991, 1994; Macdonald et al. 1995; Brand et al. 1996; reviewed by Roessler and Muenke 2001). To date, mutations in four genes, *SHH*, *ZIC2*, *PTC*, and *SIX3* have been shown to cause HPE in humans (see Table 5). However, other genetic loci for HPE have been identified showing that either essential down- or up-stream components of these pathways or other pathways remain to be found.

2.2.1 Development of the Prechordal Plate

Tissue manipulation experiments have shown that the prechordal plate (PCP), the "head organiser region", which directly underlies the developing forebrain, is essential for normal dorsal-ventral patterning of the forebrain and subsequent appropriate development of the midline of the face (Fig. 3; reviewed by Kiecker and Niehrs 2001). In the mouse, the PCP is derived from the posterior epiblast during early gastrulation and head process during late gastrulation (Tam and Meier 1982; Lawson et al. 1991; Beddington 1994; Sulik et al. 1994) whilst in amphibians and zebrafish the prechordal mesoderm is formed from the gastrula (or Spemann's) organiser and the deep hypoblast cells of the embryonic shield, respectively (Bradley et al. 1996; Foley et al. 1997; Pera and Kessel 1997; Zoltewicz and Gerhart 1997; Schneider and Mercola 1999; Gritsman et al. 2000).

The prechordal mesoderm arises following simultaneous inhibition of both BMP and Wnt signalling by the secreted factor, Cerberus (Glinka et al. 1997, 1998; Piccolo et al. 1999). Loss of BMP inhibition, in this case by gene-inactivation of the Bmp antagonists *noggin* and *chordin*, causes forebrain defects and cyclopia in mice (Bachiller et al. 2000). In addition, prechordal mesoderm induction and/or maintenance requires high levels of activin and nodal signalling (Green et al. 1992; Thisse and Thisse 1999;

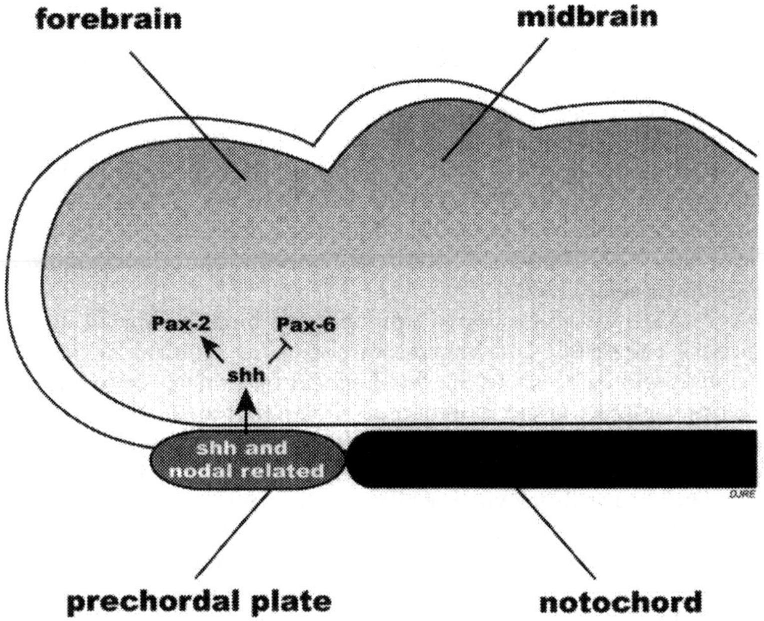

Fig. 3. Schematic showing control of midline patterning of the forebrain by the prechordal plate in an early embryo. *Sonic hedgehog* (*shh*) and nodal-related proteins are initially expressed by the prechordal plate, which lies rostral to the developing notochord. These signals subsequently promote further expression of shh, this time within the neuroectoderm, which in turn down-regulates *Pax6* expression (a lateral marker) whilst inducing/maintaining *Pax2* expression (a midline marker)

Agius et al. 2000; Gritsman et al. 2000). Loss of function of *nodal* in mice and zebrafish, as in the *sqt* and *cyc* mutants, results in cyclopia (Hatta et al. 1991; Zhou et al. 1993; Conlon et al. 1994; Thisse et al. 1994; Feldman et al. 1998; Rebagliati et al. 1998; Sampath et al. 1998). *Sqt* and *cyc* are expressed in mesoendodermal precursors. *Cyc* is also later transiently expressed in the PCP and neuroectoderm (Feldman et al. 1998; Rebagliati et al. 1998; Sampath et al. 1998). Fate labelling studies in the *cyc/sqt* double mutants have shown that the shield containing the premesendodermal cells does not form and these cells are redirected to a neural fate (Feldman et al. 2000). Similarly, loss of function of nodal (and activin) signalling by injection of the antagonistic members of the TGFβ family, lefty-1 (antivin) or -2, into zebrafish embryos prevents mesoendoderm development, resulting in cyclopia (Bisgrove et al. 1999; Meno et al. 1999; Thisse and Thisse 1999).

Oep, an EGF-related protein and cofactor of nodal, is expressed in the organiser, PCP, and the forebrain. *Oep* mutants have a similar phenotype to the *cyc/sqt* double mutants (Schier et al. 1997; Zhang et al. 1998; Gritsman et al. 1999; Kiecker et al. 2000). Oep has also been proposed to inhibit Bmp signalling pathways (Kiecker et al. 2000). Thus, it is possible that excess Bmp activity, due to mutation in *oep*, also contributes to the mesoderm defect in addition to the loss of nodal activity. Other studies have also shown that oep is required cell autonomously to maintain the PCP (Schier et al. 1997). Hence in *oep* mutants, *cyc* is initially expressed in the PCP but later its expression declines (Sampath et al. 1998).

Furthermore, loss of function of the intracellular factor *smad2*, which mediates nodal and activin signalling, can result in a cyclopean phenotype. Following nodal activation of the TGFβ receptor, smad2 is phosphorylated and translocated to the nucleus where it acts as a transcription factor. Homozygous *smad2* mutants exhibit severe defects, including only transient mesoderm development, and the embryos die too early for analysis of its role in midline patterning (Nomura and Li 1998; Waldrip et al. 1998; Weinstein et al. 1998; Heyer et al. 1999). However, chimeric embryos made from *smad2*-deficient and wild-type cells do develop and exhibit a cyclopean phenotype (Heyer et al. 1999). Similarly, heterozygote mice for both *nodal* and *smad2* are cyclopic (Nomura and Li 1998).

Mutation in TGIF, a homeodomain protein that binds to Smad2 and acts as a transcriptional co-repressor, in humans can result in hypotelorism together with midline clefts, nasal defects, and the development of a single incisor (Wotton et al. 1999a,b; Gripp et al. 2000). Four mutations in TGIF have been identified in the amino-terminal PLDLS domain, in the DNA binding domain, and the smad binding domain in the C-terminus (Gripp et al. 2000). The C-terminal domain recruits the transcriptional repressor, histone deacetylase, whilst the N-terminus acts independently of histone deacetylase and recruits another transcriptional repressor, carboxyl-terminus binding protein, into a transcriptional complex. The mutation in the homeodomain of TGIF affects DNA binding, whereas the other three affect the ability of TGIF to act as a transcriptional repressor. In the latter cases, the mutations appear to result in an increase in TGFβ signalling pathway (Melhuish and Wotton 2000). At first glance, this seems paradoxical because as discussed above, high levels of nodal and activin signalling mediated by smad2 are important for specification of the midline. However, mutation of TGIF will also affect other signalling pathways that may contribute to the holoprosencephalic phenotype. For example, TGIF normally binds to RXR sites in promoters and thus competes with the transcription factor RXR, a sterol protein that binds to and mediates retinoid signalling (Bertolino et al. 1995). Hence, loss of function of TGIF may increase retinoid signalling which has been associated with cyclopia (de Wals et al. 1991; Kalter 1992; Sulik et al. 1995).

As movement of the prechordal mesodermal cells is required, factors that control or prevent cell movement may also affect midline development. In zebrafish, there are a few mutants with convergent-extension movement defects including *silverblick* (*slb*) and the *knypek/trilobite* double mutant (Marlow et al. 1998). *Slb* is due to a mutation in *Wnt11*, which is expressed in the mesoderm. The *slb* mutant has abnormal gastrulation movements which are associated with midline defects ranging from weak hypotelorism, where the eyes are closer together, to full fusion, i.e. cyclopia (Heisenberg and Nusslein-Volhard 1997; Heisenberg et al. 2000). Similar results have been obtained by overexpression of a dominant-negative Wnt11 protein (Tada and Smith 2000). Furthermore, overexpression of the Wnt antagonist, crescent, but not of the related factor, Frzb, blocks movement of the prechordal mesoderm precursors, also resulting in a cyclopean phenotype (Pera and De Robertis 2000). Eph/ephrinB signalling is also required for cell movement (Chan et al. 2001). Teratogenic doses of alcohol in fish result in cyclopia and this has, in part, been attributed to the prevention of cell movement during formation of the PCP. This may also explain why weak holoprosencephalic phenotypes occur in Fetal Alcohol Syndrome in humans (Cohen and Sulik 1992; Johnston and Bronksy 1991; Blader and Strähle 1998).

2.2.2
Patterning the Neural Plate

Signals from the prechordal plate are needed for the development of the ventral diencephalon (hypothalamus), which divides the eye field into two (Fig. 3; Dixon and Kintner, 1989; Pera and Kessel, 1997; Camus et al. 2000). If the prechordal plate/mesoderm (or archenteron roof in amphibians) or its precursors are removed surgically, the optic field is not divided and a single eye is formed (Li et al. 1997a; Pera and Kessel 1997; Varga et al. 1999; Saúde et al. 2000). This phenomenon occurs naturally in lower chordates, the ascidians, such as *Urochordata* and *Branchiostoma* (cephalochordata), which lack the prechordal plate and hence have one centralised eye. There are two schools of thought on how division of the eye field occurs. The first is that the eye field arises from an equipotent region of neural tissue, initially characterised by the expression of *Pax6*, together with the expression of other genes such as the *Tbx2* homologue in *Xenopus* (Li et al. 1997a). Thus, the prechordal plate down-regulates the expression of *Pax6* and *Tbx2* in the neuroectoderm whilst inducing markers of the ventral diencephalon (Li et al. 1997a). The second is that migration of posterior neural cells anteriorly divides the developing eye field (Dale et al. 1997; Varga et al. 1999). This morphogenetic movement requires the prechordal mesoderm. Therefore in the *cyc* mutant, which lacks the prechordal mesoderm, the neuroectodermal cells do not migrate and subsequently undergo apoptosis (Varga et al. 1999). The apparent conflicts between the two models may be related to the timing of the experiments, and it is likely that cell movement is required first which is followed by fine tuning from the underlying PCP.

Several candidate signals for midline specification from the PCP have been identified and include Shh, Bmp7, and the nodal-related genes. The PCP signal can be mimicked, at least partially, by Shh, which is initially expressed in the PCP and is later expressed in the medial neural plate (Krauss et al. 1993; Barth and Wilson 1995; Macdonald et al. 1995; Marti et al. 1995; Shimamura et al. 1995). Loss of Shh signalling by gene-inactivation of *Shh* in mice or by blocking the Shh signalling pathway results in a cyclopean phenotype (Chiang et al. 1996; Hammerschmidt et al. 1996; Coventry et al. 1998; Incardona et al. 1998). The heterozygote phenotype in humans is characterised by a wide range of phenotypes ranging from cyclopia to a single maxillary incisor as being the sole abnormality. Many carriers exhibit no defects at all whilst in others facial defects occur without brain malformations, showing that the brain and facial defects are not necessarily linked (Belloni et al. 1996; Roessler et al. 1996, 1997; Nanni et al. 1999, 2001; Odent et al. 1999).

Overexpression studies have shown that Shh differentially regulates neuroectodermal gene expression, as does the prechordal plate. Thus Shh ventralises the neuroectoderm and promotes the expression of *Pax2*, a marker of the optic stalk, whilst repressing the expression of *Pax6*, a marker of retinal cells (Fig. 3; Krauss et al. 1993; Barth and Wilson, 1995; Ekker et al. 1995; Macdonald et al. 1995; Chiang et al. 1996; Li et al. 1997a; Pera and Kessel 1997). However, it is currently unclear whether Shh signalling is necessary in the prechordal mesoderm, neuroectoderm, or both. *Shh* expression is down-regulated in the forebrain of *cyc* zebrafish mutants and *smad2* mutant chimeric mice, suggesting it is a target of nodal signalling (Sampath et al. 1998; Heyer et al. 1999). Furthermore, the neuroectodermal expression of *Shh* can be induced in direct response to nodal signalling, again suggesting a linear relationship (Müller et al. 2000).

Mutations in the hedgehog receptor, *patched*, have also been linked to holoprosencephalic phenotypes in humans whilst mutations in *GLI3*, an intracellular repressor of Shh signalling, result in Greig cephalopolysyndactyl syndrome characterised by an expansion of the forebrain and upper face, known as hypertelorism (Vortkamp et al. 1991; Ming et al. 1998). Similarly, the *Gli3* mouse mutant has an enlarged maxillary region whilst loss of function of *Gli2*, a downstream mediator of Shh signalling, results in microcephaly in addition to other craniofacial defects. These abnormalities include effects on the maxillary incisors or the presence of a single central incisor; the latter is characteristic of HPE (Mo et al. 1997).

Cholesterol is also essential for Shh function. Following autocatalytical processing of the Shh molecule into the N and C terminal protein products, a cholesterol moiety becomes attached to the N terminal functional fragment. The association with cholesterol is essential for normal Shh signalling, and one of its functions is thought to localise the peptide to the cell surface. Hence defects in cholesterol metabolism such as occurs in Smith-Lemli-Optiz syndrome, which is due to mutations in *7-dehydrocholesterol reductase*, an enzyme that catalyses the final steps of cholesterol biosynthesis, can result in overlapping phenotypes to those that occur following mutation of *SHH*. Indeed, mild forms of holoprosencephaly occur in 5% of Smith-Lemli-Optiz patients (Kelley 1998; reviewed by Battaile and Steiner 2000). Megalin (also called Heymann nephritis antigen), which belongs to the low-density lipoprotein (LDL) receptor family, is expressed in the apical surfaces of the neuroepithelium and has been proposed to mediate cholesterol uptake from the amniotic fluid prior to the establishment of circulation (Willnow et al. 1996). Therefore, megalin function may also be required for Shh signalling. Consistent with this possibility, gene-inactivation of *megalin* in mice results in a holoprosencephalic phenotype, which includes upper face abnormalities (Willnow et al. 1996).

Shh is unable to rescue the eye defect in *cyc* mutants, despite the fact its neuroectodermal expression is activated in response to nodal signalling. Similarly, Shh cannot compensate for the loss of Dkk1, suggesting that Shh does not totally mediate the effect of cyc or Dkk1 signalling and that a number of parallel pathways exist to specify the midline (Hammerschmidt et al. 1996; Strähle et al. 1997; Kazanskaya et al. 2000). Indeed, part of the neuroectodermal response pathway also involves *cyc* and its cofactor *oep*, and the transcription factors, *zic2* and *six3* (Hatta et al. 1991; Brown et al. 1998, 2001). Studies of the *oep* zebrafish mutants have shown that in addition to the mesodermal defects, the mutant neuroectodermal midline cells are defective and are therefore unable to respond appropriately when grafted into a wild-type environment (Hatta et al. 1991, 1994; Strähle et al. 1997). Mutations in the *SIX3* gene, which is expressed in anterior neural plate, in humans results in holoprosencephaly (Wallis et al. 1999). Four distinct mutations in the homeodomain, resulting in either missense translations or deletions, have been identified (Wallis et al. 1999).

Zic2 is a zinc finger transcription factor which is homologous to Gli proteins and can bind, although with less affinity, to the same target sites in DNA (Aruga et al. 1999; Mizugishi et al. 2001). Studies in cell lines and knock-out mice have shown that Zics either potentiate or inhibit the effect of Glis in a cell-specific manner (Aruga et al. 1999). Additionally they modify transcription through other promoter sites (Mizugishi et al. 2001). Deficiency in *Zic2* function in mice can result in holoprosencephaly due to failure of the cerebral cortex to separate and aplasia of the forebrain, in addition to spina bifida and limb defects (Nagai et al. 2000). Mutation of *ZIC2* in humans results in a holoprosencephalic brain defect with no severe facial anomalies, again illustrating that the facial and brain defects can be uncoupled (Brown et al. 1998, 2001). A relatively small-scale genetic screen has indicated that mutations in *ZIC2* account for 3%–4% of HPE patients, with just over half of these being insertion mutations resulting in the expansion of the polyalanine tracts in the C-terminus. The latter mutations are thought to act as dominant-negatives and also occur in other transcription factors such as CBFA1, resulting in the bone disorder cleidocranial dysplasia, and in the PABP2 gene, resulting in oculopharnygeal muscular dystrophy (Mundlos et al. 1997; Brais et al. 1998).

Whilst the ventral axis of the developing brain is specified by Shh, the dorsal axis of the developing brain is maintained by antagonistic signals from Bmps (-2, -4, -5, -6, and -7), which are expressed in the dorsal forebrain (Furuta et al. 1997; Golden et al. 1999). Thus application of Bmps to the developing brain *in vitro* or *in vivo* results in apoptosis and decreased proliferation of the ventral tissue.

The neural crest is also essential for normal midline patterning of the head. Ablation experiments have shown that the neural crest exerts a trophic effect on the developing forebrain, and in its absence the forebrain undergoes extensive apoptosis with the resulting cyclopean phenotype (Etchevers et al. 1999). Finally, once the facial primordia are established, modulation of outgrowth of the frontonasal mass (medial nasal processes) may also result in hypo- or hypertelorism. Studies in the chick have shown that overexpression of Shh, which is expressed in a discrete area of the facial epithelium, expands the width of the frontonasal mass whilst conversely, inhibition of Shh signalling results in hypotelorism (Hu and Helms 1999).

3 Fate and Roles of the Neural Crest, Mesoderm, and Epithelium

3.1 Neural Crest

The derivatives of the neural crest in avians, amphibians, and mammals has been determined by a combination of *in vivo* studies including radioactive thymidine tracing, labelling with the lipophilic dye, DiI, retroviral labelling, chimaeric grafting studies, and more recently transgenic approaches where the progeny of the cranial neural crest cells has been followed by marking the neural crest cells with Lac-Z (Noden 1978a,b, 1983a,b, 1986, 1988; Serbedzija et al. 1992; Poelmann and Gittenberger-de-Groot 1999; Yamauchi et al. 1999; Chai et al. 2000; Jiang et al. 2000). These studies have revealed that the cranial neural crest (CNC), as is the trunk neural crest, is pluripotent, contributing to the central and peripheral nervous system, pigment cells, and connective tissue components. However, in contrast to trunk neural crest, CNC also gives rise to smooth muscle cells and hard tissues such as the facial skeleton and components of the teeth (Fig. 4; Table 2; Le Lièvre and Le Douarin 1975; Noden 1978a,b, 1983a,b, 1988, 1991a,b; Beall and Rosenquist 1990; Couly et al. 1992, 1993,

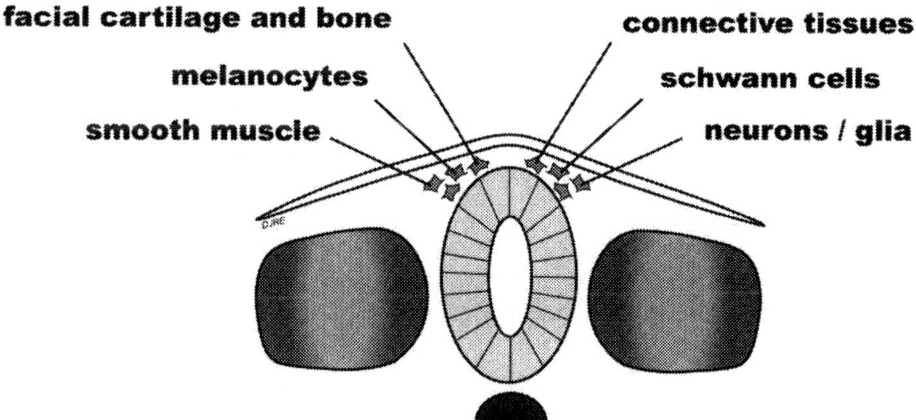

Fig. 4. A diagram showing all the possible derivatives of cranial neural crest cells. This diagram does not imply that neural crest precursors are committed to a particular fate within or as they leave the developing brain. When and where this specification occurs is still not known

Table 2. Fate of the mesoderm and neural crest derivatives arising from different positions along the rostral–caudal axis of the mouse embryo (from Trainor and Tam 1995; Trainor et al. 1994; Osumi-Yamashita et al. 1994, 1996)

Origin of neural crest	Derivative/destination	Origin of mesoderm	Derivative/destination
Forebrain		Sm I	fnm
(Stage 3–10 somites)	fnm, periocular mesenchyme, mostly rostral and ventral		Periocular mesenchyme, mostly rostral and ventral
			Endothelium of blood vessels around optic invagination
Rostral midbrain			
(Stage 3–7 somites)	fnm, periocular mesenchyme	Sm II	Periocular mesenchyme, dorsal and caudal
(Stage 3–6 somites)	Mx and Md arch		
(Stage 8–10 somites)	Close to neural tube		Mx primordia
Caudal midbrain		Sm III	Periocular mesenchyme
(Stage 3–7 somites)	Rostral and ventral periotic mesenchyme		Endothelial tissue lateral to hb
(Stage 3–6 somites)	Mx and Md arch		Branchial arch artery
(Stage 7–10 somites)	Facioacoustic ganglion		Anterior cardinal vein
			Connective tissue around truncus arteriosus
Rostral hindbrain		Dorsal	Mx primordia
	Mx and Md arch		Proximal Md arch
	Trigeminal ganglion		
		Ventral	Mx and Md, distal regions
Middle hindbrain		Sm IV Dorsal	Rostral periotic mesenchyme
	Hyoid arch		Cervical mesenchyme
	Facioacoustic ganglion		
	Rostral and ventral Periotic mesenchyme	Ventral	Hyoid arch
	Cervical mesenchyme		
		Sm V	Cervical mesenchyme
			Rostroventral periotic mesenchyme
		Dorsal	Proximal hyoid arch
		Ventral	Distal hyoid arch
Caudal hindbrain		Sm VI	3rd arch
	3rd arch		Caudal periotic mesenchyme
	Little in 2nd arch		Cervical mesenchyme
	Glossopharyngeal ganglion		
	Caudal periotic mesenchyme	Sm VII	3rd and 4th arches
			Caudal and dorsal periotic mesenchyme
			Cervical mesenchyme

Fnm = frontonasal mass; hb = hindbrain; Md = Mandibular primordia; Mx = maxillary primordia; Sm = somitomere

1995; Le Douarin et al. 1993; Imai et al. 1996; Köntges and Lumsden 1996; Chai et al. 2000). Until recently in higher vertebrates, this skeletogenic ability appears to be unique to the CNC, and hence trunk neural crest cells when grafted into the cranial environment cannot contribute to hard tissue structures, although they are capable of forming other mesectodermal derivatives (Nakamura and Ayer-le Lievre 1982). However, a recent report has shown that chick trunk neural crest cells can give rise to skeletal tissue under the appropriate conditions (McGonnell and Graham, 2002, Curr. Biol. 12: 767–771). In contrast, cranial neural crest cells retain their ability to form cartilage when transplanted into the trunk, yielding small cartilage nodules in the kidney (Le Douarin and Teillet 1974). Likewise, trunk neural crest cells cannot contribute to the smooth muscle walls of blood vessels or the meninges when grafted into the developing head (Le Douarin et al. 1977; Etchevers et al. 1999).

The cranial neural crest also contributes to the development of the aorticopulmonary septum, which separates the aorta and pulmonary trunk, and truncal septum, which separates the aortic and pulmonary semilunar valves (Kirby et al. 1983; Bockman et al. 1989; Noden 1991b; Fukiishi and Morriss-Kay 1992). This "cardiac" crest arises from the posterior hindbrain and the neural tube overlying the most anterior three somites. If this crest is ablated in the chick, defects in cardiac and artery development occur which include persistent truncus arteriosus, where the aorticopulmonary septum does not form, interruption of the aortic arch, and dextroposed aorta (Bockman et al. 1987; Nishibatake et al. 1987; Kirby and Waldo 1995).

In addition to the crucial contribution to craniofacial and cardiac structures, the neural crest cells are also needed to maintain and remodel the arch arteries and for survival of the forebrain, where the neural crest contributes to the pachymenix and leptomeninx of the meninges (Johnston 1966; Le Lièvre and Le Douarin 1975; Bockman et al. 1987; Couly and Le Douarin 1987; Noden 1991b; Couly et al. 1993; Waldo et al. 1996; Etchevers et al. 1999, 2001; Yamauchi et al. 1999). In the absence of neural crest, the forebrain undergoes apoptosis as the mesoderm, which contributes to the dura mater in the more caudal regions of the brain, cannot substitute for the loss of the neural crest (Etchevers et al. 1999). Hence, it has been proposed that the evolution of neural crest allowed the expansion of the primitive brain and the development of the forebrain (Etchevers et al. 1999, 2001). Similarly, the neural crest contributes to the thymus gland, where it promotes proliferation and survival of the endodermal cells, which will also contribute to the thymus (Bockman and Kirby 1984).

The histological and fate mapping studies have revealed differences between the timing of neural crest generation in mammals and avians. In the chick, cranial neural crest cells arise immediately after the neural folds have fused in comparison to the mouse and rat, where cranial neural crest cells are formed and migrate well before neural tube closure. Furthermore, the order of neural crest migration is different. In the chick, emigration of the cranial neural crest starts at around the 6- to 7-somite stage in the midbrain region, proceeds mainly in a rostral to caudal sequence, and is almost complete by the 16-somite stage (Anderson and Meier 1981; Tosney 1982; Lumsden et al. 1991; Sechrist et al. 1993). In contrast, in the mouse, emigration first starts at the head fold stage from the anterior hindbrain. This is followed by migration from the midbrain, then caudal hindbrain, and finally the forebrain (Nichols 1981, 1986; Chan and Tam 1988; Serbedzija et al. 1992). Migration ceases approximately 9 h after initiation in the mouse. Thus, it is complete by the 11-somite stage from the rostral hindbrain, 14 somites from the midbrain and caudal hindbrain, and 16 somite

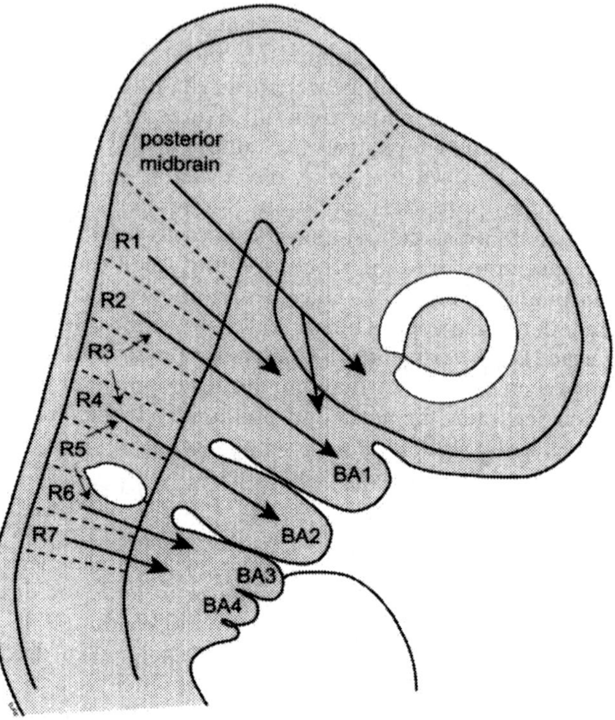

Fig. 5. Patterns of cranial neural crest cell migration. Neural crest cells are segregated into migratory streams based on their position of origin. Cells arising from the posterior midbrain and rhombomeres 1 (*R1*) and 2 (*R2*) populate the first branchial arch (*BA1*), whilst cells from rhombomere 4 (*R4*) migrate into the second arch. Neural crest cells from rhombomeres 6 and 7 (*R6* and *7*) mainly populate the more caudal arches. Surviving cells from rhombomeres 3 and 5 make smaller contributions and are directed into the first and second and the second and third arches, respectively. See Sect. 4.2 for further details

stage in the forebrain. The relative order of migration from the rat is different yet again (Tan and Morriss-Kay 1985). There are also subtle differences in the origin and migration routes of cranial neural crest cells, but despite these slight variations, overall the origin and destinations are remarkably similar between the different species.

In rats, mice, and avians, the hindbrain crest migrates ventrolaterally under the ectoderm and is segregated into three streams (Fig. 5). The most rostral stream (trigeminal) arises from rhombomeres 1 and 2, which contributes to the first arch (maxillary and mandibular primordia) and the trigeminal ganglion. The middle stream (hyoid) arises from r4 and contributes to the second or hyoid arch and vestibulocochlear ganglion. In the chick, surviving cells from r3 and r5 have been shown to mainly give rise to the second arch skeleton plus some neuronal derivatives (Köntges and Lumsden 1996; Couly et al. 1998). The most caudal population (post-otic) is formed from rhombomeres 6 to 7 and gives rise to components of the third, fourth, and sixth branchial arches together with the nodose ganglion (Noden 1975,

1978a; D'Amico-Martel and Noden 1983, Lumsden et al. 1991). This crest population in chicks has also recently been shown to contribute to the connective tissue of the clavicle (McGonnell et al. 2001). The midbrain crest migrates as a sheet of cells, underneath the ectoderm in the chick but through the mesenchyme in mice and rats, giving rise to the distal components of the first arch, periocular region, and frontonasal mass. The prosencephalon neural crest cells migrate ventrally to form the part of the upper head (Johnston 1966; Noden 1975, 1978a; Tam and Morris-Kay 1985; Lumsden et al. 1991; Fukisshi and Morriss-Kay 1992; Serbedzija et al. 1992; Matsuo et al. 1993; Morriss-Kay et al. 1993; Osumi-Yamashita et al. 1994, 1996; Trainor and Tam 1995).

The final destination of the neural crest cells is dependent on the timing of emigration. In the hindbrain, there is a progressive ventral to dorsal sequence. The first cells to migrate populate the branchial arches, whereas the last progeny do not migrate extensively and contribute to the ganglia lying adjacent to the neural tube (Lumsden et al. 1991; Serbedzija et al. 1992; Osumi-Yamashita et al. 1994, 1996). At the 3-somite stage in mice, the entire mesencephalon contributes to both the frontonasal mass and first branchial arch, whereas by the 5- to 10 somite stages, only neural crest cells from the anterior mesencephalon give rise to the frontonasal mass whilst the more posterior region specifically contributes to the trigeminal ganglion of the first branchial arch (Osumi-Yamashita et al. 1994).

In chicks there is another migratory cell population that contributes to the head. These cells arise from the ventral hindbrain at approximately stage 16, after neural crest generation on the dorsal surface has been completed, and exit the neural tube via the trigeminal ganglion attachment. Along with the dorsal cells, they contribute to cartilage, nerve, and connective tissue components of the head and heart (Lunn et al. 1987; Loring et al. 1988; Sohal et al. 1996, 1999a,b, 1998; Bockman and Sohal 1998).

3.2 Mesoderm

The cranial paraxial mesoderm and prechordal plate give rise to muscle, skeletal structures, endothelial cells, pericytes, and meninges of the brain, with the exception of the pachymenix and leptomeninx around the forebrain (Le Lièvre and Le Douarin 1975; Peault et al. 1983; Pardanaud et al. 1987; Coffin and Poole 1988; Noden 1982, 1983a, 1988; Poole and Coffin 1988; Couly et al. 1992, 1995). During development, the cranial paraxial mesoderm is thought to be transiently segregated into seven loose aggregations of cells, somitomeres (Sm), which contribute to arches 1–4, but not 6 (Meier 1982; Meier and Tam 1982; Tam and Meier 1982; Jacobson 1988; Tam and Trainor 1994; Trainor et al. 1994). Sm 1 underlies the prosencephalon, Sm II and III are found lateral to the mesencephalon, Sm IV is adjacent to the metencephalon, whilst Sm V–VII underlie the myelencephalon (Fig. 6; Noden 1983a; Couly et al. 1992; Trainor et al. 1994). Fate labelling studies in chicks and mice have shown that the rostral somitomeres give rise to the majority of extraocular muscles and the frontal bone, whilst those more caudally give rise to the precursors of the jaw muscles and the parietal bone plus some of the bones of the skull base. Endothelial cells migrate widely from all somitomeres (Noden 1991a; Couly et al. 1995). In the mouse, each branchial

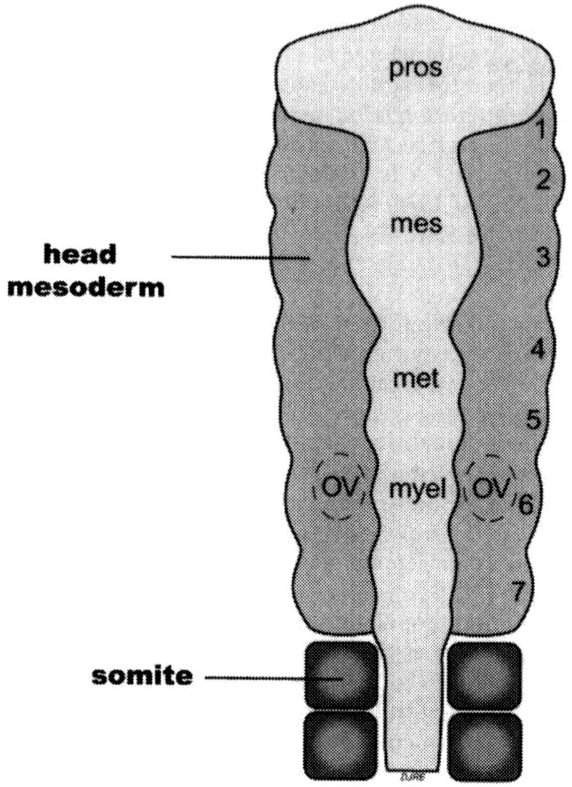

Fig. 6. Diagram of the proposed sites of somitomeres within the cranial paraxial mesoderm of avian embryos. *pros*, Prosencephalon; *mes*, mesencephalon; *met*, metancephalon; *myel*, myelencephalon; *OV*, otic vesicle. The long-term fate of each somitomere, and the location of its derivatives, compared to the location of neural crest derivatives from the neural tube at the same rostral-caudal axis in mice is shown in Table 2. (Based on the diagrams of Noden 1983a)

arch receives a contribution from two somitomeres (Tam and Trainor 1994; Trainor et al. 1994).

These fate-labelling studies have also revealed that the myogenic derivatives, in general, stay in register along the rostro-caudal axis, with neural crest that arises in the adjacent neural tube (Table 2; Noden 1982; Osumi-Yamashita et al. 1994; Trainor et al. 1994; Hacker and Guthrie 1998). The one main exception is the lateral rectus/pyramidalis complex, which arises adjacent to the border of the mesencephalon and metencephalon but migrates to the pericocular mesenchyme (Wahl et al. 1994). With the exception of the branchial arches, there is extensive mixing and there is no clear neural crest/mesoderm interface (Fig. 7; Noden 1986; Trainor and Tam 1995). In branchial arches 1 and 2, the neural crest cells encapsulate a mesodermal core, whereas in branchial arches 3 and 4, the neural crest cells surround the artery and the mesoderm is found at the periphery. Later, within the mesodermal core of branchial arches 1 and 2, clusters of neural crest cells are found, which have been proposed to segregate the myogenic cells from the other tissues derived from the somitomere (Noden 1988; Trainor and Tam 1995).

The endothelial cells are the first lineage to become segregated from the mesodermal derivatives, as shown by the expression of the VEGF receptor, Quek, in the chick (Eichmann et al. 1993, 1996; Couly et al. 1995). This occurs by the head fold stage when,

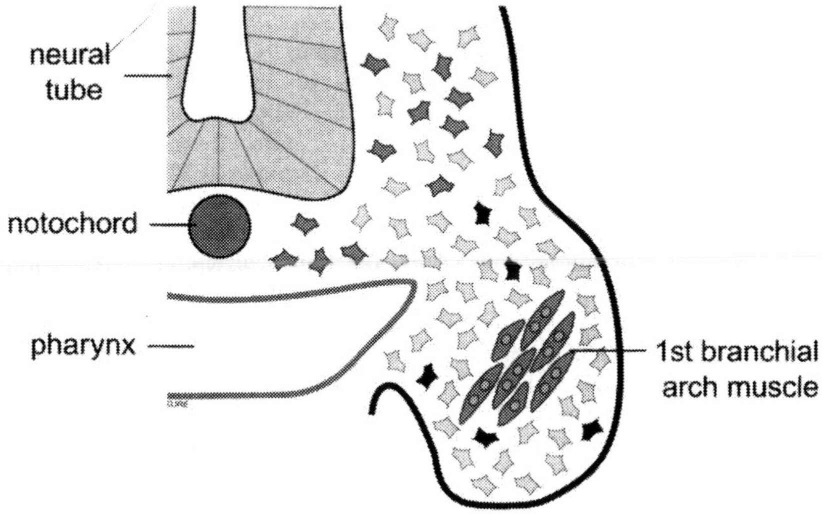

Fig. 7. Diagram of the movement of neural crest cells (*light grey*) and myogenic cells (*dark grey*) into the first branchial arch in the avian embryo. In arches 1 and 2, neural crest cells initially surround the developing arch muscles, separating them from the surface ectoderm, before investing the muscle masses to form the muscle connective tissues. Angiogenic cells (*black*) are also interspersed within the arch mesenchyme. (Based on Wahl and Noden 2001)

with the exception of the prechordal plate, 50% of the mesodermal cells express Quek. In the prechordal plate, the percentage is much lower, reflecting its more limited angiogenic capacity (Couly et al. 1995). In contrast, the onset of myogenesis and skeletal differentiation occurs much later (see Sects. 6 and 7). In the chick, differentiated endothelial cells are first present in the cephalic mesoderm by embryonic day (E) 2 and these cells first invade the neuroepithelium at E3, whereas they populate the muscles much later at E7. This difference in timing presumably reflects the greater growth of the brain and hence the need for vascularisation (Pardanaud et al. 1987; Poole and Coffin 1988; Coffin and Poole 1988).

Using chick-quail transplantations and retroviral lineage tracing, the skeletal derivatives of the paraxial mesoderm have been found to include the lateral walls of the skull in the occipital and petrous regions and the floor of the braincase from the occipital level to that of the basisphenoid region (Noden 1978a,b; 1984, 1988). The origins of the cranial vault, however, is more controversial, with Noden (1978, 1988) demonstrating that both the parietal and frontal bones (except for the rostral supraorbital region) are also derived from the mesoderm, whilst Couly et al. (1993) claim that these bones are entirely neural crest-derived. Recent work from the lab of Morriss-Kay, looking at the mammalian skull, suggests that there are also differences in the origins of the avian and mammalian skull vaults (Jiang et al. 2002). Using X-gal staining and DiI labelling, these workers showed that the mammalian frontal bones are of neural crest origin whilst the parietal bone and interparietal bones (not evident in the avian) are mesoderm-derived. To reconcile these differences, it has been suggested that in the reptilian ancestors of modern birds, the frontal and parietal bones fused together (Jiang et al. 2002), explaining the dual neural crest/mesodermal origin of the avian frontal bones as demonstrated by Noden (1978a,b,1988). In this way, the parietal bones in birds therefore probably equate to the mammalian interparietal bones. The mesodermal/neural crest contribution to other skeletal structures of the head also varies in different species. For example, the pleurosphenoid bone, which forms the lateral wall of the brain case, is formed from mesoderm in reptiles and avians, whilst the "functionally homologous" bone in mammals, the alisphenoid bone, is formed from neural crest. Neural crest transplantation studies in the chick have revealed that both neural crest and mesoderm are equivalent in their ability to respond to skeletogenic inducing factors (Fyfe and Hall 1979; Noden 1983b; Schneider 1999). Thus, neural crest cells can become incorporated into skeletal structures when transplanted ectopically into the mesoderm, showing that environmental signals are key to patterning the skeletal structures. Indeed a few bones in the head, such as the otic capsule, basisphenoid, and frontal bone in the chick as discussed above, are derived from both mesodermal and neural crest contributions (Le Lièvre 1978; Noden 1982, 1983a). This also suggests plasticity within the system where the neural crest and mesodermal contributions to the bones has been/can be easily modified during development and evolution.

The mesoderm plays a number of crucial roles in craniofacial development. First, the prechordal mesoderm determines midline patterning of the forebrain (see Sect. 2.2). Second, mesodermal signals can specify neural tube fate co-ordinating anterior-posterior patterning of the hindbrain (see Sect. 4.2). Third, mesodermal signals induce the development of taste buds in the endoderm and in synergy with neural signals, direct otic induction (Ladher et al. 2000a; Barlow 2001; Francis-West et al. 2002). Fourth, mesodermal signals are involved in neural crest generation and possible migration (see Sects. 4.1 and 4.3). Finally, mesodermal signals maintain Hox gene expression in the migrating neural crest. As will be discussed later, this will have a direct effect on craniofacial patterning (see Sect. 5.1).

3.3 Epithelium

Fate mapping studies in avians and mice have shown that the anterior neural ridge gives rise to the upper facial ectoderm overlying the frontonasal mass and premaxillary bone, the roof of the oral cavity, olfactory placodes and their derivatives, together with the glandular cords of the adenohypophysis, which is derived from Rathke's pouch (Couly et al. 1985, 1987; Osumi-Yamashita et al. 1994). The ectoderm in other regions of the face arises from the ectoderm adjacent to the neural tube, and it stays in register with the neural crest derivatives arising from the adjacent neural tube. For example, the ectoderm adjacent to the rostral hindbrain forms the ectoderm over the first arch, oral cavity, and tongue (Couly and Le Douarin 1990). Thus, the cranial neural crest and overlying ectoderm expand continuously together (D'Amico-Martel and Noden 1983; Couly and Le Douarin 1990). This ectoderm also gives rise to the placodes, including the lens, otic, and epibranchial placodes, which form part of the peripheral nervous system. For reviews of placode development and function, see Baker and Bronner-Fraser 2001 and Francis-West et al. 2002. The endoderm lining the branchial pouches gives rise to the thymus, parathyroid, and thyroid glands (Le Douarin et al. 1976).

The epithelium has a role in neural tube closure (see Sect. 2.1) and the induction of some cranial placodes, such as the epibranchial placodes, and the neural crest together with its later differentiation into skeletogenic and odontogenic structures (see Sects. 4.1 and 6.1 and Thesleff and Sharpe 1997 and Jernvall and Thesleff 2000 for reviews of the role of the epithelium during odontogenic induction and patterning). In addition, epithelial signals control outgrowth and patterning of the facial primordia and have also been proposed to control neural crest migration (see Sects. 4.3 and 5.1).

4 Neural Crest Development

4.1 Generation

Neural crest arises at the interface of the ectoderm and neural tissue via inductive interactions between the two tissues (Fig. 8; Rosenquist 1981; Moury and Jacobson 1990; Selleck and Bronner-Fraser 1995; Mancilla and Mayor 1996). In fact, placement of the ectoderm and the neural plate together in *in vivo* or *in vitro* recombination assays can induce neural crest formation with neural crest cells forming from both the ectoderm and neural tissue (reviewed by Selleck and Bronner-Fraser 1996; LaBonne and Bronner-Fraser 1999). Most of these studies have been carried out in the trunk, but the tissue interactions that control cranial neural crest formation are likely to be similar if not identical. However, as recently shown by the studies of the effect of slug overexpression in the developing trunk and cranial regions, the molecular mechanisms of neural crest generation are not exactly identical (del Barrio and Nieto 2002). These molecular differences are also emphasised by the differential expression of genes such as the hyaluronan receptor, *CD44*, and the transcription factor, *Id2*, which are restricted to the cranial neural crest (Martinsen and Bronner-Fraser 1998; Corbel et al. 2000; also see Sect. 4.3 and the review by Nieto 2001). In chicks, the neural/epithelial boundary, which gives rise to the neural crest precursors, is set up by antagonistic interactions between Bmp and Fgf signalling in conjunction with a signal from the midline of the neural plate (Pera et al. 1999; Streit and Stern 1999). At this early stage, the presumptive neural crest cells are not specified and are competent to produce neural crest in addition to neuronal and ectodermal derivatives (Selleck and Bronner-Fraser 1995). Later, they become restricted to either a neural crest or neuronal fate.

However, neural crest generation is not simply an interaction between the epidermal and neural tissues, and the mesoderm has been implicated to have a role. Studies in the axolotl have shown that neural versus neural crest fate is determined by the underlying archenteron roof (endomesoderm) (Raven and Kloos 1945). The medial region can induce both neural tissue and neural crest whilst the lateral region can only induce neural crest in the appropriate responding tissue. Similar studies in chicks and *Xenopus* have shown that the non-axial mesoderm induces the neural border (Bang et al. 1997; Bonstein et al. 1998; Marchant et al. 1998). Thus, if this mesoderm is ablated in *Xenopus*, neural crest cells are not formed. Fgfs and Wnts are thought to be involved in this mesodermal signalling (Fig. 8; Muhr et al. 1997, 1999; reviewed by Basch et al. 2000; Bronner-Fraser 2002). For further details about the role of Wnt signalling during

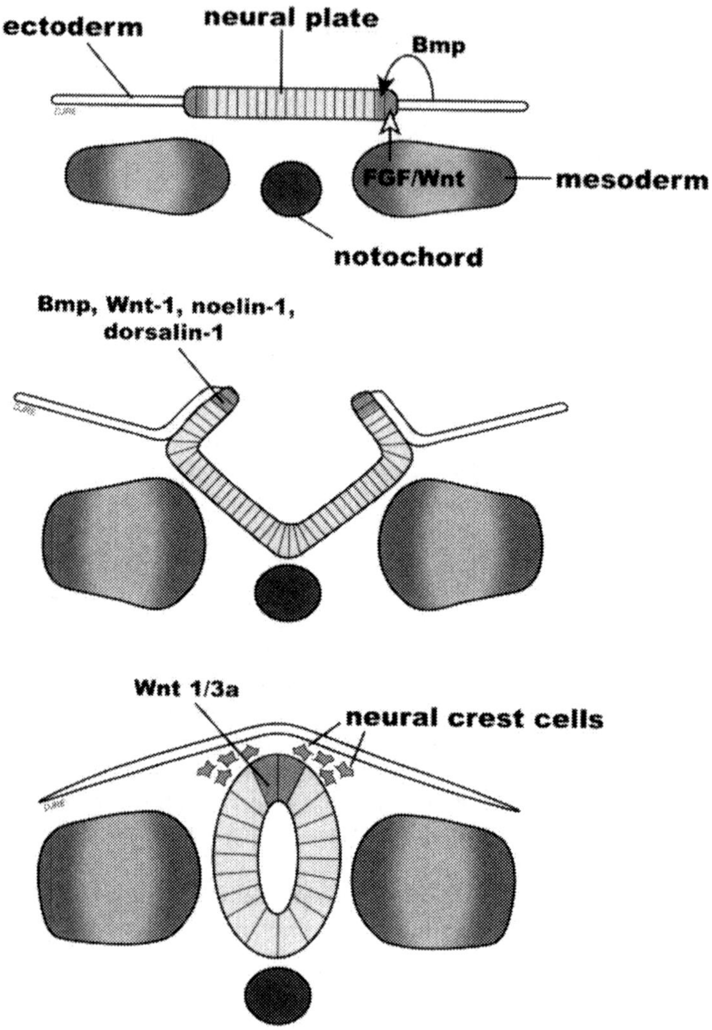

Fig. 8. Schematic diagram of the induction of neural crest cell development within the dorsal neural tube. Bmp, Fgf, and Wnt family members are involved in the induction of the neural plate border, the presumptive neural folds (*dark grey shading*). Continued expression of Bmp in the ectoderm flanking the developing neural tube leads to the induction of neural crest formation within the neural folds. Factors expressed within the neural folds and which control neural crest generation include Wnt-1, noelin-1, Bmp, and dorsalin-1. See Sect. 4.1 for further details

neural crest development, see the review by Wu an colleagues (Wu et al. 2003 Trends in neuroscience 26: 40–45).

Bmps, in contrast, are thought to signal from the ectoderm to induce neural crest formation. In the chick and mouse, *Bmps* (*-2*, *-4* and/or *-7*) are expressed in the ectoderm flanking the neural tube (Fig. 8; reviewed by Selleck and Bronner-Fraser 1996; LaBonne and Bronner-Fraser 1999). *Bmp7* is also expressed in the paraxial mesoderm and the presumptive neural crest cells (Arkell and Beddington 1997; Solloway and Robertson 1999). Transgenic expression of the Bmp antagonist, noggin, which blocks the function of Bmp2, -4 and -7 in the cranial premigratory and migratory neural crest cells results in the loss of neural crest progeny similar to that seen in *Bmp2* knockout mice (Kanzler et al. 2000). This suggests that Bmps are the inductive signals for neural crest formation and that Bmp2 may be the key molecule. Indeed in chicks, the rostral limit of neural crest generation corresponds to the rostral limit of *Bmp* expression (Muhr et al. 1997). However, forebrain neural plate cells do not form neural crest even when exposed to Bmps, suggesting that there are other differences between neural crest-producing and -non-producing regions in the head. Similarly, in *swirl* zebrafish mutants, which have a mutation in the *Bmp2* gene, there are defects in neural crest production (Nguyen et al. 1998). Recent studies in the chick have shown that the expression of *Bmp4* in the non-neural ectoderm is controlled by notch signalling (Endo et al. 2002).

Neuroectodermal response genes, which are expressed in the presumptive neural crest, include the secreted factors, the TGFβ member, *dorsalin*, *noelin-1*, *Wnt-1*, and *Bmps* (Fig. 8). The transcriptional mediators include slug and snail, sox9, Id2, together with members of the FoxD and Zic families (Nieto et al. 1994; Martinsen and Bronner-Fraser 1998; Sefton et al. 1998; Hromas et al. 1999; Barembaum et al. 2000; Deardorff et al. 2001; Pohl and Knöchel 2001; Tan et al. 2001; Spokony et al. 2002). The role of these genes has been inferred by overexpression or functional ablation studies, and all have been proposed to have a role in the establishment of the premigratory neural crest precursors. In particular noelin-1, a novel secreted factor, has been proposed to confer competence to the neural tube to produce neural crest (Barembaum et al. 2000). Overexpression of either Zic2 (or the related factors, Zic 1, -3, and -5), slug, Id2, noelin-1, or dorsalin, have been shown to result in hyperplasia of the neural crest (Nagai et al. 1997; Nakata et al. 1997, 1998, 2000; Brewster et al. 1998; Kuo et al. 1998). In contrast, antisense oligo ablation of *sox9* in *Xenopus* embryos or *slug* expression in developing chick and *Xenopus* embryos prevents neural crest development (Nieto et al. 1994; Carl et al. 1999; Spokony et al. 2002). In the mouse, *slug* is not expressed until after neural crest formation, and it is proposed that the related factor, snail, has an analogous role (Smith et al. 1992; Jiang et al. 1998b; Sefton et al. 1998).

The role of zic2 in neural crest generation is further indirectly supported by the *Zic2* "knockdown" mouse, in which the levels of *zic2* transcripts are only 20% of wild-type levels. In these mice, there is a delay in neurulation with a consequent, although possibly secondary, defect in neural crest development (Nagai et al. 2000). Similarly, levels of FoxD3 (also known as fkd6 in zebrafish, CWH-3 in chickens, genesis/HFH-2 in mice) may regulate neural crest generation. Overexpression of *FoxD3*, which is expressed in the presumptive neural crest prior to *slug* expression, down-regulates the expression of *slug*, *snail*, *cadherin-11*, and *FoxD3* itself, and hence prevents neural crest generation. A similar phenotype results from overexpression of the related factor, *XBF-2*, which is expressed transiently in neural crest (Mariani and Harland 1998; Gómez-Skarmeta et al. 1999).

Following induction, the presumptive neural crest cells change shape and alter their cell and matrix interactions, resulting in delamination (Duband et al. 1995). In the trunk neural crest, there is a switch in the expression profile of cadherins, membrane-bound proteins that mediate cell to cell interactions. Concomitant with neural crest formation, the expression of N-cadherin, E-cadherin, and cadherin 6B is down-regulated whilst that of cadherin-7 is switched on (Revel and Brown 1976; Akitaya and Bronner-Fraser 1992; Nakagawa and Takeichi 1995, 1998). This change in cadherin profile is crucial for the segregation of neural crest precursors from neural cells, and if N-cadherin expression is maintained, neural crest formation is prevented (Nakagawa and Takeichi 1998). In the cranial neural crest, N-cadherin is also down-regulated in the premigratory neural crest whilst cadherin-11 is switched on. The α_4 integrin receptor and RhoB, which change cell adhesion and/or cell shape, are also required for neural crest development (Kil et al. 1998; Liu and Jessel 1998). Inhibition studies of RhoB, which is expressed in the dorsal lips of the neural folds, have pin-pointed precisely when its function is required, showing that it is essential for delamination, but not for specification or migration, of the neural crest (Liu and Jessel 1998). Snail and slug also appear to be essential for this epithelial-mesenchymal transformation and delamination, as suggested by the ability of snail to down-regulate *E-cadherin* expression decreasing cell adhesion in epithelial cells and to up-regulate the number of cells expressing RhoB in the neural tube (Cano et al. 2000; del Barrio and Nieto 2002). The neural crest cells subsequently migrate through the ectodermal/neuroectodermal junction, which lacks a basal lamina (Erickson and Perris 1993; Miner et al. 1998).

4.2 Segregation of Hindbrain Neural Crest

In vertebrates, the hindbrain neural crest is clearly segregated into three main streams; those that populate the first branchial arch, the second arch, and the more caudal arches (Fig. 5). This segregation of the neural crest populations is important, particularly between the neural crest cells that will populate the first and second arches, i.e. R1/2 and R4 streams, as these neural crest populations control patterning of the respective arches. As will be discussed in Sect. 5.1, neural crest cells emigrating from rhombomere 4 express *Hoxa2* which controls patterning of the second or hyoid arch (Fig. 9). Likewise, "first arch neural crest," i.e. that arising from rhombomeres 1/2 and the midbrain controls patterning of the proximal first arch structures. Thus, segregation of these neural crest populations is essential for normal head development. Neural crest cells arising from rhombomeres 6–8 also express distinct Hox codes, which may be important in determining third/ fourth arch derivatives.

This segregation is in part achieved by the division of the hindbrain into rhombomeres, which provide the template for the Hox code expressed by the migrating neural crest and organisation of the cranial ganglia. As shown in Fig. 9, each rhombomere expresses a characteristic Hox code which is controlled by vertical signals from the somites and planar signals from within the neuroectoderm (Grapin-Botton et al. 1995, 1997; Itasaki et al. 1996; Taneja et al. 1996; Gould et al. 1998). One factor involved is retinoic acid (RA), which is needed for the formation of the rhombomere boundaries and rhombomere specification (Gould et al. 1998; Niederreither et al. 2000; reviewed by Gavalas and Krumlauf 2000). Indeed, there is a gradient of RA activity across the developing hindbrain. During gastrulation and neurulation, RA is produced by the posterior mesoderm, including that adjacent to R6/7. In contrast, in the anterior of the embryo, RA is degraded by Cyp26 (also known as 450RAI), a P450 enzyme, which is expressed in the presumptive mid- and forebrain (Swindell et al.

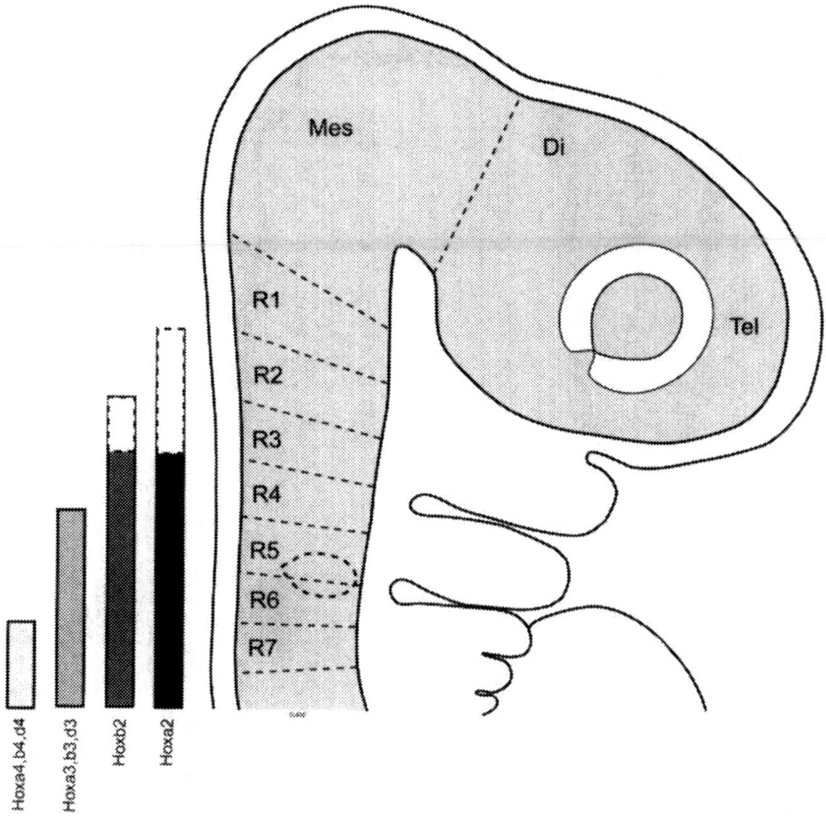

Fig. 9. Schematic diagram of the characteristic expression of Hox genes within the developing hindbrain. Where the bars are shaded, the Hox gene is expressed in both the neural tube and its neural crest derivatives. Where the bar is shown by a dotted outline, the Hox gene is expressed in the neural tube only. Therefore, Hoxa4, -b4 and d4 expression is restricted to the level of rhombomere 7 (R7), whilst Hoxa3, -b3 and d3 are expressed in rhombomeres 5–7. Hoxb2 is expressed up to the level of R3, and Hoxa2 to the level of R2. Hox gene expression patterns are similar within the neural cells migrating out from each rhombomere, except that Hoxa2 is not expressed by neural crest cells from rhombomeres 2 and 3 and Hoxb2 is not expressed by neural crest cells emanating from R3. The down-regulation of Hoxa2 expression in the first arch neural crest is essential for normal first arch development (see Sect. 5.1 for more details). *Mes*, mesencephalon; *Di*, diencephalon; *Tel*, telencephalon

1999). Later, *Cyp26* is also expressed in rhombomere 2 (Swindell et al. 1999). Thus, a gradient of RA controls patterning of the rostral-caudal axis of the hindbrain, presumably accounting for the particular sensitivity of the hindbrain to changes in RA signalling (reviewed by Maden 1999). Recent studies have also implicated FGF signalling in the specification of rhombomeres 3 and 5 (Marin and Charnay 2000).

In chicks, neural crest migration starts after the formation of the distinct rhombomeres, and therefore each neural crest population arises from a segregated region of the brain (Lumsden and Keynes 1989; Fraser et al. 1990; Birgbauer and Fraser 1994). In contrast, in mammals the developing hindbrain is initially divided into prorhom-

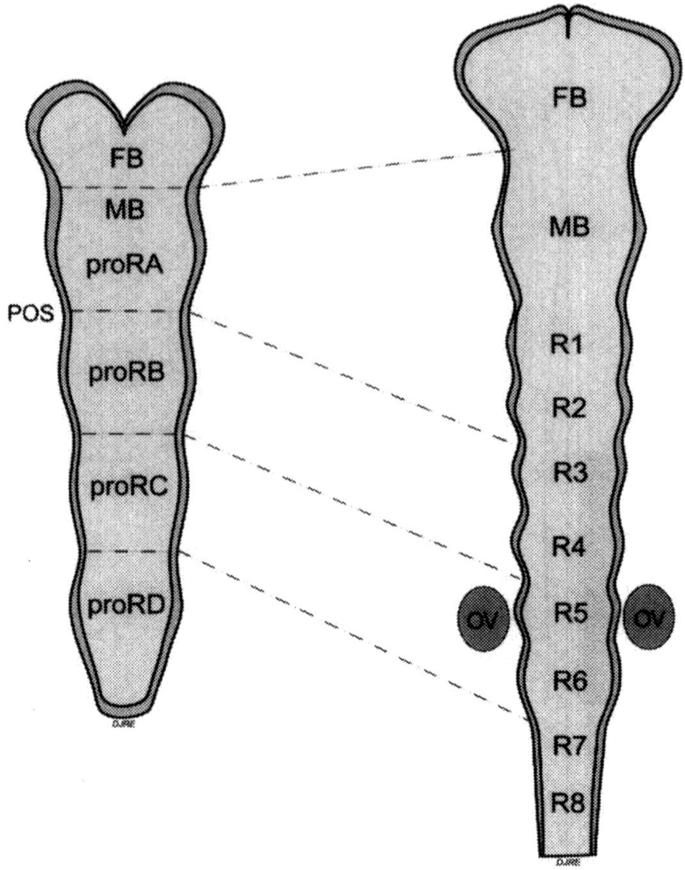

Fig. 10. Illustration of the initial segmentation of the developing mouse hindbrain into prorhombomeres (*proRs*). These segments subsequently divide forming the definitive rhombomeres (*Rs*). *FB*, forebrain; *MB*, midbrain; *POS*, pre-otic sulcus; *OV*, otic vesicle. (Based on Osumi-Yamashita et al. 1996, 1997)

bomeres: proRhA, proRhB, proRhC, and proRD, which subsequently divide into the definitive rhombomeres (Fig. 10; Osumi-Yamashita et al. 1996; Trainor and Tam 1995). In mice, neural crest cell migration begins at the 4-somite stage (E8.5) and is almost complete by 7- to 14- somite stage; i.e. the majority of crest has migrated before the formation of the distinctive rhombomeres (Nichols 1981, 1986; Chan and Tam 1988; Serbedzija et al. 1992). Despite this, however, fate labelling studies have shown that these prorhombomeres also seem to restrict cell lineage (Trainor and Tam 1995; Osumi-Yamashita et al. 1996). This is consistent with the onset of discrete domains of *Hox* expression prior to the development of the rhombomere boundaries, showing that there are already intrinsic differences between the presumptive rhombomeres before overt changes in morphology (Graham et al. 1989; Gould et al. 1998).

In chicks and mice, there are no migratory neural crest streams arising from R3 and R5, although these rhombomeres are capable and do generate neural crest. The R3 and R5 neural crest cells that are generated and that survive, migrate either anteriorly or posteriorly within or along the dorsal surface of the neural tube to join the adjacent neural crest streams (Anderson and Meier 1981; Sechrist et al. 1993; Birgbauer et al. 1995; Köntges and Lumsden 1996). Once they enter the adjacent streams, R3 and R5 neural crest cells do not maintain the *Hox* gene expression characteristic of their rhombomere of origin. Thus, the expression of *Hoxb2* and *Hoxb3* has been shown to be lost in R3 and R5 progeny, respectively. However, they may not completely change their molecular fate. For example, in mice some neural crest cells from R5, which migrate caudally, have been shown to maintain the expression of *Krox20*, characteristic of R5 but not R6 derivatives, at least initially during their migration (Sham et al. 1993). Long-term fate mapping studies in chicks by the generation of quail-chick chimeras or DiI labelling studies have shown that these neural crest cells migrate into the arches contributing to the skeletal structures or give rise to the later wave of neural crest derivatives, the ganglion (Köntges and Lumsden 1996; Couly et al. 1998).

In chicks and mice, these distinct streams of neural crest appear to be generated, at least in part, by a "paraxial mesoderm exclusion zone" adjacent to rhombomeres 3 and 5 (Fig. 11). Following grafts of R4 into the R3 position, the neural crest cells did not migrate perpendicularly from the rhombomere, as may be anticipated in the

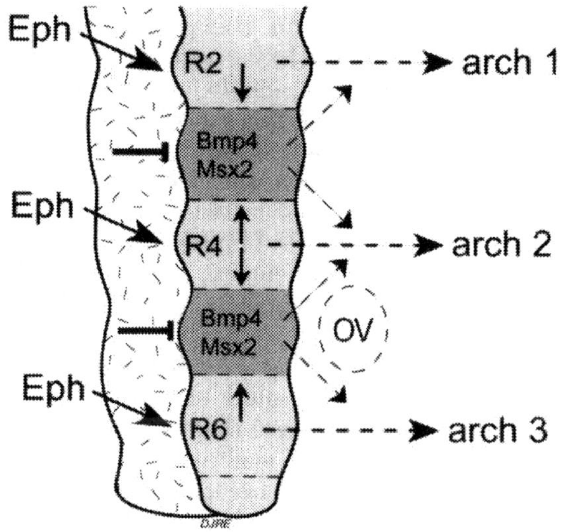

Fig. 11. Diagram of the signals that have been implicated in the segregation of neural crest cells from rhombomeres 2 and 4 in mice and chicks. This segregation is important as it will control patterning of the lower jaw (see Sect. 5.1) and appears to be controlled by a "paraxial mesoderm exclusion zone", which is inhibitory for neural crest migration and/or generation, adjacent to R3 and R5. This is in part achieved by ephrin signalling within the neuroectoderm, neural crest cells themselves, and the ectoderm. In chicks, the later neural crest progeny of R3 and R5 undergo apoptosis, further segregating the R2 and R4 neural crest populations. This apoptosis is induced by the even-numbered rhombomeres (indicated by arrows), which maintain the expression of the apoptotic signal, Bmp4, and its downstream transcriptional mediator, Msx2. The surviving neural crest cells from R3 and R5 are still able to migrate, however, and move either anteriorly or posteriorly to join adjacent neural crest streams. See Sect. 4.2 for further details. *OV*, otic vesicle

absence of external inhibitory signals, but instead migrate anteriorly and posteriorly to join the adjacent neural crest streams (Niederlander and Lumsden 1996; Farlie et al. 1999). Conversely, Kuratani and Eichele (1993) observed that small pieces of R3 when grafted into R6 induced a neural crest free zone in the adjacent host tissue. This suggests that R3 induces the expression of factors that prevent neural crest generation, survival, and/or neural crest migration in the mesoderm.

This idea is supported by molecular studies in mice and chicks. For example, in the *ErbB4* knockout mouse, neural crest cells from R4 migrate into the mesoderm adjacent to R3 and join the R2 stream, down-regulating *Hoxb2* expression in the process (Gassmann et al. 1995; Golding et al. 2000). Grafts of mutant neural crest cells into a wild-type environment has shown that this mismigration is due to an environmental defect and is not due to changes in the neural crest cells themselves (Golding et al. 2000). *ErbB4* is expressed in R3 and R5 in both mice and chicks, and therefore a signal from the neuroectoderm must control the properties of the adjacent mesoderm (Gassmann et al. 1995; Dixon and Lumsden 1999). In mice, this repulsive effect is mediated by chondroitin sulphate binding molecules as chondroitinase treatment removes the inhibitory effect of “R3 mesoderm” (Golding et al. 1999). In chicks, *semaphorin III* (or *collapsin-1*), has also been shown to be expressed adjacent to R5 and to be repulsive for neural crest cells, which express its receptor, *neuropilin 1* (Eickholt et al. 1999).

Differential ephrin signalling, which as will be discussed later controls cell sorting, may also control migration (Fig. 11). For example, the ectodermal expression of *ephrinB2* correlates with regions avoided by the neural crest, suggesting it is repulsive. Consistent with this, cranial neural crest cells do not migrate in distinct streams but are scattered in the *ephrinB2* null mouse. This neural crest defect, in part, contributes to the ultimate hypoplasia of the second arch and the vascular defects observed in the *ephrinB2* mutant mice (Wang et al. 1998; Adams et al. 1999, 2001). There may also be positive attractive or guidance cues, which localise the neural crest streams. For example, in the developing mouse, *ephrinA5* expression is excluded from the surface ectoderm adjacent to rhombomere 3 (Flenniken et al. 1996; Robinson et al. 1997). Therefore, as in the trunk, neural crest migration may be regulated by permissive and inhibitory signals.

In avians but not in mice, this sculpting is further refined as some of the neural crest cells arising from rhombomeres 3 and 5 undergo programmed cell death or apoptosis (Fig. 11). Cell death within these rhombomeres is greatest at the 13–15 somite stages, after the production of the majority of neural crest (Jeffs et al. 1992). Random small pockets of cell death also occur in other regions of the hindbrain including R2, R6 together with the R1/2 and R2/3 boundaries (Jeffs et al. 1992; Graham et al. 1993; Farlie et al. 1999; Ellies et al. 2000). These localised pockets of cell death are also seen in the mouse hindbrain and may be neural crest cells that have failed to attach to extracellular matrix which, at least for trunk neural crest, is essential for cell survival (Meredith et al. 1993; Frisch and Francis 1994; Trainor et al. 2002b).

In vitro and *in vivo* studies in chicks have shown that the signals for apoptosis arise from the even-numbered rhombomeres in an anterior-posterior wave (Graham et al. 1993, 1994; Ellies et al. 2000). Thus, placement of an impermeable barrier between R2 and R3 prevents apoptosis of the R3 neural crest. Similarly, removal of the even-numbered rhombomeres either by ablation *in vivo* or culture of R3 alone *in vitro* results in the survival of the neural crest cells that would normally have undergone apoptosis

(Graham et al. 1993, 1994; Ellies et al. 2000). In contrast, when R3 is cultured with an even-numbered rhombomere *in vitro*, the R3 neural crest cells die (Graham et al. 1993, 1994; Ellies et al. 2000).

Three key factors have been implicated in this process – Bmp4, its presumed downstream transcriptional target Msx2, and Sfrp-2, a secreted Wnt antagonist. *Bmp4*, together with its receptors *Bmpr1A* and *BmprII*, and *Msx2* are initially expressed throughout the developing hindbrain (Graham et al. 1993 1994; Farlie et al. 1999; Smith and Graham 2001). Later, co-incident with the localisation of apoptosis, *Bmp4* and *Msx2* expression become predominantly localised to R3 and R5 (Fig. 11). Addition of recombinant Bmp4 to isolated cultures of R3, which under these conditions do not undergo apoptosis, induces cell death in the neural crest cells, suggesting that Bmp4 is the mediator of cell death. In contrast, R4 neural crest cells, which are adjacent to the *Bmp4*-expressing rhombomeres, are resistant to the apoptotic effects of Bmp4 as they express the secreted Bmp antagonist, *noggin* (Smith and Graham 2001). Msx2 appears to be the immediate transducer of the Bmp4 signal. First, as in other developmental systems *Msx2* expression is induced by Bmp4, and second, retroviral misexpression of Msx2 in the developing chick hindbrain induces apoptosis in the even-numbered rhombomeres (Takahashi et al. 1998).

Expression of *Bmp4* may be regulated by Wnt signalling. The secreted Wnt antagonist, *Sfrp-2*, is initially expressed throughout the developing hindbrain and is later down-regulated in rhombomeres 3 and 5 at and prior to the onset of apoptosis, respectively (Ladher et al. 2000b; Ellies et al. 2000). Overexpression of Sfrp-2 prevents cell death in R3, and this is associated with a down-regulation of *Bmp4* (Ellies et al. 2000). Conversely, loss-of-Sfrp-2 function, by antisense RNA strategies, in R4 induces cell death (Ellies et al. 2000).

In *Xenopus*, the hindbrain neural crest cells are not separated by mesodermal exclusion zones or localised foci of apoptosis, and there are distinct mechanisms of segregation. Thus, these populations remain together and do not separate until they enter the arches (Sadaghiani and Thiebaud 1987; Smith et al. 1997). These distinct populations of neural crest cells are kept apart by the differential expression of ephrins, membrane-associated proteins, which mediate either repulsive or interactive cell-cell interactions. The ephrin family has been divided into A and B classes. The former usually interact with EphA receptors whilst the latter usually signal through the EphB receptors. Both classes are expressed in neural crest. The second arch neural crest and mesoderm express *Ephrin-B2*, whilst the third arch expresses the *EphA4* and *EphB1* receptors (Jones et al. 1995; Smith et al. 1997). Following misexpression of dominant-negative EphA4 or EphB1 receptors, the third arch crest cells mismigrate anteriorly and posteriorly (Smith et al. 1997). Misexpression of ephrin-B2 also results in neural crest mismigration (Smith et al. 1997). Likewise, the EphA2 receptor, which is expressed in the second and third arch, keeps the third and fourth neural crest populations apart (Helbling et al. 1998). Similarly, as discussed above, ephrins and their receptors also have a role in segregating neural crest streams in chicks and mammals (for review see Robinson et al. 1997).

4.3 Migration

Following specification and delamination, the neural crest cells migrate along specific pathways to their final destination (also see review by Krull 2001). The majority of the initial wave of neural crest progeny will enter the branchial arches and the upper head whilst the later wave contributes to the sensory ganglion (see Sect. 3.1). The former migrate through the mesoderm whereas the latter migrate ventrally along the surface of the neural tube (Tosney 1982). Neural crest migration is not just a case of passive displacement following their generation, and the two processes have been uncoupled. Studies in both the chick and mouse have shown that when neural crest migration is inhibited, such as following blocking FGF2 or serotonin function, neural crest cells accumulate within the neural tube (Moiseiwitsch and Lauder 1995; Kubota and Ito 2000).

As the cranial neural crest cells migrate they express the cell surface molecules, β_1 and α integrin subunits, and N-CAM together with cadherin-11 (Bronner-Fraser et al. 1992; Borchers et al. 2001). During migration, they encounter the integrin ligands, fibronectin and laminin, together with tenascin and heparin sulfate proteoglycan in the cranial mesenchyme (Krotoski et al. 1986; Duband and Thiery 1987; Bronner-Fraser 1988). These matrix molecules may be expressed by the neural crest cells themselves or by cells along their pathway. An example of the former is the expression of *fibronectin*. Experiments in which blocking antibodies or dominant-negative constructs have been injected into the cranial mesenchyme have shown that β_1 and $\alpha 4$ subunits of integrin, cadherin-11, N-CAM, galactosyltransferase, fibronectin, tenascin, and a laminin-heparan sulfate proteoglycan complex are all required for migration of at least a subpopulation of cranial neural crest cells (Boucaut et al. 1984; Bronner-Fraser 1985, 1986, 1987; Poole and Thiery 1986; Runyan et al. 1986; Bronner-Fraser and Lallier 1988; Bronner-Fraser et al. 1992; Lallier et al. 1992; Kil et al. 1998; Borchers et al. 2001). Migration is also controlled by proteases, such as ADAM 13, which degrade the extracellular matrix (Alfandari et al. 2001).

The molecular mechanisms of trunk versus cranial neural crest migration appear to be distinct. For example, trunk neural crest cells do not require tenascin, laminin-HSPE complex, or fibronectin for migration and require different integrins (Le Douarin and Teillet 1974; Le Lièvre et al. 1980; Bronner-Fraser 1985, 1986, 1987; Bronner-Fraser and Lallier 1988; Lallier and Bronner-Fraser 1992, 1993; Lallier et al. 1992). Furthermore, trunk, but not cranial, neural crest cells can bind to collagens I and IV whilst cranial, but not trunk, neural crest cells produce fibronectin *in vitro* culture (Newgreen and Thiery 1980). However, despite these differences, the general principles are likely to be the same involving the differential expression of permissive versus inhibitory substrates (reviewed by Perris 1997; also see Sect. 4.2 for examples of inhibitory interactions that control segregation of neural crest populations).

DiI labeling studies have suggested that the migration of neural crest cells is random (Kulesa and Fraser 1998). Other studies have suggested that chemoattractive signals may exist. Fgf2, which is expressed in the presumptive first arch, has been shown to be a chemoattractant for midbrain neural crest cells (Kubota and Ito 2000). A recent study has also implicated FGF signalling in the control of neural crest migration into the second arch (Trokovic et al. 2003; Genes Dev. 17: 141–153). In addition, *scatter factor*, which is expressed in the branchial arch mesenchyme, may play a role. Scatter factor has been shown to be at least in part responsible for the projection of the cranial motor neurons, which shares many parallels with neural crest migration (Caton et al. 2000). Other indirect evidence for a chemoattractant includes the changes in neural crest migration observed following rhombomere rotation in the chick (Sechrist et al. 1994). In this case, the neural crest cells have been noted to migrate towards ectopic otic vesicles, suggesting that the latter maybe chemoattractive, at least initially. At later stages, the otic vesicle expresses semaphorin III, which is repulsive for neural crest cells (Eickholt et al. 1999). In addition, following transplantation of R6 to R4, a subpopulation of these cells have been shown to migrate towards the third arch, suggesting that they may be an attraction or segregation signal within the third arch mesenchyme (Nieto et al. 1995; Saldivar et al. 1996). Finally, retinoid excess in E7.5 mice converts R2/3 to R4/5 identity (Marshall et al. 1992). The resulting "R2/3" crest cells migrate ectopically into the second arch and toward the presumptive second arch derivative, the acoustic-facial ganglion, suggesting that they have acquired second arch neural crest identity and are migrating toward a chemoattractive source (Lee et al. 1995). In all cases, these effects could be linked to the differential expression of ephrins, which control cell sorting and guidance into the arches, as was discussed in Sect. 4.2.

The neurotransmitter serotonin (5-hydroxytryptamine) is also needed for migration (Moiseiwitsch and Lauder 1995; Choi et al. 1997). Thus, treatment of developing embryos with antagonists such as ritanserin or the 5-HT2B-like receptor, methysergide, *in vivo* and *in vitro* blocks migration (Lauder and Zimmerman 1988; Moiseiwitsch and Lauder 1995; Choi et al. 1997). Following antagonist treatment the neural crest cells stay at the dorsal surface of the neural tube, yielding embryos with head defects including hypoplastic arches

Subpopulations of neural crest cells from the midbrain, rhombomeres 1 and 7, have been shown to move as chains in contact with each other (Kulesa and Fraser 1998). Therefore, gap-junctions, which allow the passage of ions, second messengers, and small metabolites between cells may facilitate this migration co-ordinating movement and later, possibly differentiation. Following loss of function of *α_1-connexin* (also known as connexin 43), a component of gap-junctions, there are conotruncal heart malformations which are at least in part due to defects in the cardiac neural crest (Ewart et al. 1997; Huang et al. 1998; Sullivan et al. 1998). *In vitro* studies with connexin-overexpressing or connexin-deficient neural crest cells has shown overexpression of α_1-connexin increases migration whilst loss of function has the opposite effect (Lo et al. 1997; Huang et al. 1998). N-Cadherin function is also required for migration and gap-junctional communication, but it does not mediate its effect by altering levels of α_1-connexin (Xu et al. 2001).

The transcription factors AP-2, slug, snail, twist, FoxD3, and the Bmp4 transducer, smad1, are expressed in the migratory cranial neural crest cells. Of these, slug and FoxD3 have been shown to be needed for migration in addition to neural crest generation (Carl et al. 1999; LaBonne and Bronner-Fraser 2000). In the case of slug, this has been shown by loss of function experiments in *Xenopus* embryos by overexpressing dominant-negative constructs or antisense oligo strategies (Carl et al. 1999; LaBonne and Bronner-Fraser 2000). The closely related transcription factor, snail, has been shown to repress E-cadherin expression in cell lines, and hence the loss of slug function in migrating neural crest cells may increase the expression of cell adhesion molecules, preventing further migration (Cano et al. 2000). Overexpression of FoxD3, which is expressed in the presumptive and migratory neural crest cells, prevents neural crest migration. This appears to be at least in part due to changes in the environment, as wild-type neural crest cells when grafted into a FoxD3 overexpressing environment either mismigrate or don't migrate at all (Pohl and Knöchel 2001).

Another transcription factor, Pax6, is essential for normal migration of a specific population of neural crest cells – the anterior midbrain neural crest (Matsuo et al. 1993). Mutations in the *Pax6* gene are responsible for the *small eye* rat and mouse mutants (Hill et al. 1991; Matsuo et al. 1993). The heterozygotes are characterised by small eyes whilst the homozygotes lack the lens and olfactory placode together with the lateral nasal process (Matsuo et al. 1993; Fujiwara et al. 1994; Kaufman et al. 1995). Fate labelling studies in rats have shown that the anterior midbrain neural crest cells do not migrate beyond the dorsal aspect of the developing eye (Osumi-Yamashita et al. 1997). In mice, these mislocated neural crest cells may undergo apoptosis as inferred from the enlarged region of cell death in the cells overlying the forebrain at E10.5 in *small eye* mutant mice (Grindley et al. 1995, 1997). Chimeric grafting studies in which wild-type neural crest has been placed in a mutant environment has shown that the defect is due to changes in environmental signals and is not due to abnormalities in the neural crest cells themselves. Thus, neural crest cells from the *small eye* mutant migrate normally in a wild-type host whereas wild-type neural crest cells in a *small eye* mutant do not (Osumi-Yamashita et al. 1997). At the time of migration, *Pax6* expression is regionalised to the developing forebrain and to the ectoderm overlying the forebrain but not that overlying the hindbrain. As *Pax6* is not expressed in the migrating neural crest cells themselves, it may regulate the expression of secreted/membrane-bound factors that control migration in the surrounding environment (Walther and Gruss 1991; Puschel et al. 1992; Li et al. 1994; Grindley et al. 1995). Candidate molecules include the cell adhesion molecules N-CAM and L1 together with integrins α_5 and β_1, which have Pax6 binding sites in their promoters (Chalepakis et al. 1993; 1994; Holst et al. 1997; Duncan et al. 2000; reviewed by Edelman and Jones 1998). The related factor Pax3 is required for migration of cardiac neural crest, and loss-of-function is associated with an upregulation of the chondroitin sulphate molecule, versican (Epstein et al. 1993, 2000; Conway et al. 1997a,b; Henderson et al. 1997). Pax3 is also thought to be important for the generation of neural crest cells, which will also be a contributory factor to the cardiac defects seen in *Splotch* mice (Conway et al. 2000).

In vivo and *in vitro* studies have shown that excess RA inhibits neural crest migration, and in the chick, migrating cranial neural crest cells express the retinoid receptor RARβ suggesting that they are the direct targets of RA signalling (Rowe et al. 1991; Moro-Balbas et al. 1993). Indeed, RA administration prolongs the expression of N-CAM by the neural crest cells and changes the expression of matrix molecules, which will affect neural crest migration (Moro-Balbas et al. 1993; Shankar et al. 1994). These effects will in part account for the teratogenic effect of excess RA, which results in craniofacial abnormalities. However, the role of retinoids during normal migration is as yet unclear as neural crest migratory defects have not been reported in any of the *RAR* or *RXR* knock-out mice (Lohnes et al. 1994; Mendelsohn et al. 1994a,b; Luo et al. 1996; Kastner et al. 1997; Dupé et al. 1999).

4.4 Proliferation and Survival

Several factors that control survival of the migrating and post-migratory neural crest have been identified. These include serotonin, Shh, Bmps, Fgfs, endothelin-1, PDGF, and α5-integrin (Moiseiwitsch and Lauder 1995; Choi et al. 1997; Goh et al. 1997; Soriano 1997; also see Sect. 5 and Table 3). Injection of Shh blocking antibodies into the cranial mesenchyme adjacent to the hindbrain during neural crest migration results in the formation of hypoplastic arches and a reduction in the frontonasal mass (Ahlgren and Bronner-Fraser 1999). Whilst cell death is increased, proliferation is unaffected, showing that in contrast to other regions of the embryo including the facial primordia, Shh does not act as a mitogen. A recent study has shown that neural crest cell death as a result of alcohol excess is due to down-regulation of *Shh* expression. Furthermore, Shh can rescue the teratogenic effects of alcohol on cranial neural crest development (Algren et al. 2002; Proc. Natl. Acad. Sci. 99: 10476–10478). Overexpression of another secreted factor, fringe, which is expressed in the neural tube and adjacent ectoderm, can increase neural crest proliferation (Nellemann et al. 2001).

Loss of *α5-integrin* function in mice affects neural crest cell survival in the second arch whilst α4 integrin is needed for neural crest proliferation (Goh et al. 1997; Kil et al. 1998). In addition, the transcription factors AP-2 and Pax6 are required for cell survival (Grindley et al. 1995; Schorle et al. 1996; Zhang et al. 1996). Cell death in the latter may to be linked to the mismigration of the anterior midbrain neural crest cells that occurs following loss of *Pax6* function. The decrease in cell survival following loss of *AP-2* may be in part mediated by the loss of serotonin and/or α5-integrin signalling as the serotonin receptor, *5-HT2B*, and *α5-integrin* genes have AP-2 binding sites in their promoters (Zutter et al. 1997; Kambe et al. 1998).

The multifunctional protein β-catenin is also essential for neural crest survival. β-catenin is a transcriptional transducer of the Wnt1 signalling pathway and a component of the cadherin/actin cytoskeleton. Gene-inactivation of *β-catenin* in the presumptive and migratory cranial neural crest cells results in apoptosis of both neural and proximal mesenchymal derivatives of the neural crest in the first and second branchial arches together with increased apoptosis in the frontonasal mass (Brault et al. 2001).

Table 3. Mouse mutants that have craniofacial defects[a]

Gene	Expression	Phenotype	Mechanism	Reference
Transcription factors				
Homeobox-containing genes				
Alx3/4	Frontonasal mass and mandible	Mid-facial cleft – two nasal septa Cleft palate Nasal capsule hypoplastic Skull and cranial base defects Premaxilla and maxilla affected	Abnormalities arise at E10.5 Increase in levels of apoptosis in frontonasal mass Levels of proliferation are unchanged	Beverdam et al. 2001
Alx4/Cart1	Craniofacial mesenchyme	Mid-facial cleft Cleft palate Microagnathia Skull bone defect – some are smaller	Not known	Qu et al. 1999
Bapx1	Head mesenchyme Facial mesenchyme	Basal skull defects (mesodermally derived bones)	Chondrogenic differentiation prevented Apoptosis of chondrogenic precursors	Akazawa et al. 2000 Lettice et al. 1999 Tribioli and Lufkin 1999
Cart1	Cranial mesenchyme Skeleton	Microagnathia Loss of cranial vault Nasal cartilage – hyperplasia or agenesis	Skeletal differentiation?	Zhao et al. 1996
Dlx2	Migratory and post-migratory neural crest	First and second arch defects	Not known	Qiu et al. 1995
Dlx5	Neuroepithelium Surface ectoderm Facial mesenchyme Skeleton	Microagnathia Cleft lip and palate Nasal defects – hypoplastic Ectopic bones Inner ear defects Tooth defects	Loss of Gsc in mandible and frontonasal mass Msx1 and Msx2 are unchanged Levels of apoptosis are unchanged proliferation affected Nasal epithelium defective Skeletal differentiation?	Acampora et al. 1999 Depew et al. 1999
Goosecoid	Facial mesenchyme Tongue muscle	Microagnathia Aplastic nasal cavities and epithelium Pharyngeal arch muscle defects Nasal septum does not fuse with secondary palate (although they do meet) Inner ear defects Cranial base defects	Downstream of Dlx5 in nasal region	Belo et al. 1998 Rivera-Perez et al. 1995 Yamada et al. 1995 Yamada et al. 1997

Table 3. *Continued*[a]

Gene	Expression	Phenotype	Mechanism	Reference
Lhx8 (also L3 and Lhx7)	Mandibular and maxillary primordia Palatal shelves	Cleft palate	Palatal shelves are small and do not extend towards midline	Zhao et al. 1999
Msx1	Facial mesenchyme Osteogenic precursors	Microagnathia Cleft palate Skull ossification defects (enlarged anterior fontanelle) Nasal defects	Epithelial-mesenchymal interactions Delayed ossification - reduced cell proliferation of osteogenic precursors	Houzelstein et al. 1997 Satokata and Maas 1994
Msx2	Facial mesenchyme Osteogenic precursors	Persistent calvarial foramen	Delayed ossification – reduced cell proliferation of osteogenic precursors	Satokata et al. 2000
Msx1/Msx2		Loss of membrane bones		Satokata et al. 2000
Pax6	Ectoderm and neuroepithelium	Microagnathia Nasal defects Lens defects Ectopic cartilage Tooth defects	Defect in neural crest migration Change in cell adhesion?	Fujiwara et al. 1994 Kaufman et al. 1995 Matsuo et al. 1993
Pax9	Facial mesenchyme and branchial arches	Cleft palate Thymus and parathyroid agenesis	Abnormal development of palatal shelves	Peters et al. 1998
Pitx1	Mandibular primordia	Cleft palate Microagnathia Tongue is short and bifurcated	Palatal shelves do not elevate	Lanctot et al. 1999 Szeto et al. 1999
Pitx2	Oral ectoderm Cranial mesenchyme	Microagnathia Cleft palate Extraocular muscles are absent (do not form condensations) Cardiac defects	Fgf-8 expression absent Bmp4, Msx1 and 2 expression expanded Palatal shelves and oral cavity small (tongue fills cavity)	Gage et al. 1999 Lu et al. 1999b
Prx1/2	Facial mesenchyme and second arch	Microagnathia Cleft palate, mandible, and tongue Second arch defects Ectopic cartilage	Controls shh expression in overlying epithelium Reduced proliferation	Lu et al. 1999a ten Berge et al. 1998 ten Berge et al. 2001

Table 3. *Continued*[a]

Gene	Expression	Phenotype	Mechanism	Reference
Tbx1 (candidate gene for DiGeorge syndrome)	Epithelium and mesoderm	Microagnathia Cleft palate Thymus and parathyroid agenesis Aortic arch and cardiac defects Second arch defects	Not known	Jerome and Papaioannou 2001 Lindsay et al. 2001 Merscher et al. 2001
Other transcription factors				
Ap2	Neural crest and facial mesenchyme	Midline facial cleft Cleft mandible Skeletal dysmorphogenesis/agenesis (hyoid, Meckel's and mandible are normal)	Increase in apoptosis in first arch Midline cleft independent of exencephalic phenotype	Nottoli et al. 1998 Schorle et al. 1998 Zhang et al. 1998
Gli2	Facial mesenchyme Skeleton	Microagnathia – bones are absent Cleft palate Skull defects – delayed ossification Cranial base defects Inner ear defects	Palatal shelves do not elevate or elevation is delayed	Mo et al. 1997
Gli3 mutant	Facial mesenchyme Skeleton	Cleft palate Enlarged maxillary region Smaller nasal processes Skull defects	Not known	Mo et al. 1997
Hic1 (deleted in Miller-Dieker syndrome)	Ubiquitous	Microagnathia Cleft palate Cleft mandible Holoprosencephaly and cyclopia Nasal chonchae are absent Skull bones absent	Palatal shelves are too small	Carter et al. 2000
Hes-1	Neuroepithelium Cranial mesenchyme	Mid-facial cleft	Secondary to exencephaly?	Ishibashi et al. 1995
Hesx1	Forebrain Rathke's pouch	Microagnathia	Not known	Dattani et al. 1998

Table 3. *Continued*[a]

Gene	Expression	Phenotype	Mechanism	Reference
δEF1	Headfold and cephalic mesenchyme Skeleton	Microagnathia Cleft palate Hyperplasia of Meckel's cartilage Nasal septum dysplasia	Not known	Funahashi et al. 1993 Takagi et al. 1998
MFH-1	Cranial mesenchyme Skeleton	Cleft palate Many cranial bones are dysmorphic Aortic arch and heart defects	Aortic arches are not remodelled	Iida et al. 1997
Rae28 (polycomb gene)		Cleft palate Thymus and parathyroid agenesis Cardiac defects	Anterior shift in the expression of some Hox genes	Takihara et al. 1997
Ski	Neuroepithelium Migrating neural crest	Microagnathia Mid-facial cleft Cleft lip Skull bones absent	Increase in apoptosis in mesenchyme Levels of proliferation are unchanged Cells sparser in mesenchyme	Berk et al. 1997
Smad2	Ubiquitous	Cyclopia Mandible – absent or hypoplastic	Loss of PCP Shh expression lost in neuroectoderm	Heyer et al. 1999 Nomura and Li 1998 Waldrip et al. 1998 Weinstein et al. 1998
Titf2	Ectoderm and endoderm	Thyroid agenesis and ectopia Cleft palate Pulmonary malformations	Not known	De Felice et al. 1998
p63	Branchial arch epithelium	Microagnathia Cleft palate Branchial arches are hypoplastic Hyoid bone is absent	Epithelial-mesenchymal interactions Maintenance of epithelium?	Mills et al. 1999 Yang et al. 1999
Growth factors and receptors				
Activinβ-A	Facial mesenchyme	Cleft palate Tooth defects	Palatal shelves are too small	Matzuk et al. 1995b

Table 3. *Continued*[a]

Gene	Expression	Phenotype	Mechanism	Reference
Activin type IIA receptor (ALK4)		Mandible hypoplasia Cyclopia Cleft palate	Not known	Matzuk et al. 1995a
Bmp5/7	Cranial mesoderm and ectoderm	Branchial arches are smaller	Increase in the levels of apoptosis in the branchial arch ectoderm Fewer cells present in branchial arches	Solloway and Robertson 1999
Chordin/noggin (Bmp antagonists)	Migrating neural crest (noggin) Mandibular ectoderm (noggin and chordin) Meckel's cartilage (noggin)	Microagnathia Meckel's cartilage larger	Decrease in Fgf-8 expression Increase in the levels of apoptosis	Stottmann et al. 2001
Connexin 43	Migrating neural crest cells Osteoblasts	Cranial bones hypoplastic Open foramen Microagnathia	Neural crest migration defect Osteoblast defect – delayed ossification	Lecanda et al. 2000
Ephβ2 and β3	Palatal shelves	Cleft palate	Palatal shelves are too small	Adams et al. 1999 Orioli et al. 1996
EGFR	Mandibular primordia and palatal shelves Ligands (EGF and TGFα) in MEE	Narrow long snouts Impaired Meckel's cartilage development Cleft palate	Remnants of MEE persist Regulation of metalloproteinase expression	Miettinen et al. 1999
ECE-1	Cranial mesenchyme and epithelium	Lower jaw hypoplastic Thyroid and parathyroid hypoplastic Thymus ectopia External ear defects Aortic arch and heart defects	Defective epithelial-mesenchymal interactions	Yanagisawa et al. 1998b
ET-1	Epithelium and mesoderm	Lower jaw hypoplastic Cleft palate Inner and outer ear defects Tongue defects Aortic arch patterning defects	Defective epithelial-mesenchymal interactions	Kurihara et al. 1994 Yanagisawa et al. 1998a

Table 3. *Continued*[a]

Gene	Expression	Phenotype	Mechanism	Reference
ET_A receptor	Migrating and post-migratory neural crest	First and second arch hypoplasia Ear defects Heart and aortic arch defects	Loss of homeobox-gene expression Increase in the levels of apoptosis Decrease in the levels of proliferation	Clouthier et al. 1998 Clouthier et al. 2000 Yanagisawa et al. 1998a
Follistatin (activin antagonist)	Palatal shelves Olfactory mesenchyme Tongue	Cleft palate Mandible – slightly truncated Tooth defects	Palatal shelves are too small	Matzuk et al. 1995c
α5-integrin	Ubiquitous	Cleft palate Hypoplastic second arch	Palatal shelves are small and do not meet at midline Decrease in cell survival in mesenchyme and endoderm	Bader et al. 1998 Goh et al. 1997
Jagged 2 mutant	Facial epithelium	Cleft palate Tongue fused to palatal shelves Thymus differentiation defects	Palatal shelves do not elevate	Jiang et al. 1998a
PDGFRα (Patch mutant)	Migrating neural crest Facial mesenchyme Heart	Cleft frontonasal mass Cleft mandibular primordia Cleft palate Heart valve and septation defects Thymus agenesis/hypoplasia	Decrease in neural crest migration Decrease in MMP-2 and MT-MMP i.e. abnormal tissue remodelling Increase in apoptosis in mandibular primordia Abnormal matrix	Morrison-Graham et al. 1992 Robbins et al. 1999
Ryk (tyrosine kinase-related orphan receptor)	Base of oral cavity Palatal shelves Tongue	Cleft palate Microagnathia Calvarial defects Microcephaly	Interacts with Ephβ2 and β3 receptors	Halford et al. 2000
Shh	Neuroepithelium Facial epithelium	Cyclopia Craniofacial bones severely hypoplastic or absent	Decrease in cell survival and proliferation Failure to induce midline Loss of Tbx1 in branchial arches	Chiang et al. 1996
Mes mutant (patch – shh receptor mutant)	Target tissues of Shh signalling	Microagnathia Hypertelorism Domed head	Shh overactivity? (C terminal of Ptc lost)	Makino et al. 2001

Table 3. *Continued*[a]

Gene	Expression	Phenotype	Mechanism	Reference
TGFβ2	Palatal shelves Skeleton	Calvarial and mandibular bones are smaller and dysmorphic Cleft palate Cardiac defects	Palatal shelves do not elevate	Sanford et al. 1997
TGFβ3	MEE of palatal shelves	Cleft palate	Shelves meet but MEE persists and shelves do not fuse	Kaartinen et al. 1995 Proetzel et al. 1995
Wnt1/3a	Dorsal neural tube	Second arch abnormalities	Decrease in neural crest production	Ikeya et al. 1997
Wnt5a	Migratory neural crest and facial mesenchyme	Microagnathia	Decrease in cell proliferation?	Yamaguchi et al. 1999
Neurotransmitters				
GABA67	Maxillary primordia	Cleft palate	Palatal shelves elevate but do not fuse	Asada et al. 1997 Condie et al. 1997
β3 GABA receptor	Palatal shelves	Cleft palate	Decreased cell motility in palatal shelves?	Homanics et al. 1997
Matrix proteins				
α1(II) collagen		Cleft palate		Pace et al. 1997
Cho/Cho mutant (Col IIαI)		Cleft palate Short head Protruding tongue	Cleft due to defective lower jaw growth	Lavrin et al. 1997
Link protein	Skeleton	Flat face Short and domed skull	Affects endochondral bones Stabalises hyaluronan and aggrecan	Watanabe and Yamada 1999
Perlecan	Skeleton	Microagnathia Cleft palate Domed skull	Abnormal matrix	Arikawa-Hirasawa et al. 1999

Table 3. *Continued*[a]

Gene	Expression	Phenotype	Mechanism	Reference
Intracellular molecules				
Apaf1		Midline facial cleft Cleft palate	Defect secondary to exencephaly?	Cecconi et al. 1998 Honarpour et al. 2000 Yoshida et al. 1998
β-catenin	Premigratory and migratory neural crest cells	Neural crest-derived structures are hypoplastic or absent	Increase in the levels of apoptosis in post-migratory neural crest cells	Brault et al. 2001
p57kip2 (cdk inhibitor)	Palatal shelves Muscle cells of tongue	Cleft palate	Not known	Caspary et al. 1999 Zhang et al. 1997
CASK		Microagnathia Pointed snout Microcephaly Cleft palate – sex linked	Not known	Laverty and Wilson 1998 Wilson et al. 1993
CRKL (SH2-SH3-SH2 adaptor protein)	Neuroepithelium Migrating and post-migratory neural crest	Hypertelorism Thymus and parathyroid agenesis Base of skull is wider and shorter Cardiac and aortic arch defects	Neural crest generation and migration are normal Post-migratory defect	Guris et al. 2001
Fidgetin (ATPase)	Branchial arches Skeleton Inner ear	Skull bones fused Inner ear defects	Not known	Cox et al. 2000
IKK1		Cleft palate Tooth defects	Not known	Li et al. 1999
MacMARCKs		Hypotelorism	Not known	Chen et al. 1996 Wu et al. 1996
SH2 tyrosine phosphatase		Branchial arch defects	Involved in Fgf-8 signalling?	Saxton et al. 2000
Tattered mutant (Δ8-Δ7 sterol)		Microagnathia Cleft palate	Missense mutation	Derry et al. 1999

Table 3. *Continued*[a]

Gene	Expression	Phenotype	Mechanism	Reference
isomerase binding protein)		Broad domed skull Hypertelorism		
Treacle	Presumptive and post-migratory neural crest	Hypoplastic upper and lower face Nasal placode defects Palatal defects	Increase in apoptosis in premigratory (and post-migratory?) neural crest	Dixon et al. 2000
3β-hydroxysterol D7-reductase gene (cholesterol biosynthesis)		Cleft palate Lack nasal opening	Palatal shelves elevate but are too small to fuse	Wassif et al. 2001
Shroom (novel PDZ domain protein)		Facial clefting	Face defect secondary to exencephaly?	Hildebrand and Soriano 1999

[a]The mouse mutants are listed under the subheadings "Homeobox-containing genes," "Other transcription factors," "Growth factors and receptors," "Matrix proteins," and "Intracellular molecules," in alphabetical order within each section. The growth factor section also includes the growth factor antagonists. The expression domain and phenotypes reported are restricted to the craniofacial tissues for clarity. Our current knowledge about the mechanism of action of the gene and the molecular and/or cellular effect of loss of its function is given under the heading "Mechanism".

5 Growth and Patterning of the Face and Branchial Arches

5.1 Patterning of the Face and Branchial Arches

The craniofacial structures are mainly designed for feeding and protecting the brain together with the sense organs, the eye, nose, and ear. How these complex structures arose is a fundamental question for both evolutionary and developmental biologists. One possibility is that the lower jaw, for example, evolved by modification of the gill bearing arches found in jawless vertebrates (reviewed by Kimmel et al. 2001; Kurantani et al. 2001). Following the formation of the jaw apparatus such as that found in amphibians, there have been further adaptations in mammals such as the acquisition of the distinct temporomandibular joint, which has allowed full development of the mammalian auditory system.

The face is a jigsaw of several neural crest populations, which integrate to form the skeletal structures. This integration where neural crest populations form "cryptic boundaries" within a skeletal structure suggests that either one population must be instructive or that the fate of the neural crest is controlled by environmental signals (Köntges and Lumsden 1996). The latter appears to be generally true and large advances have recently been made in the identification of the tissue interactions involved, although the molecular players, particularly those controlling development of the upper jaw, remain elusive. Development of each region of the face and head appears to be controlled by a unique set of tissue interactions. The isthmus has recently been evoked to control patterning of the proximal first arch, the endoderm patterns the facial skeleton whilst the ectoderm/neuroectoderm has been proposed to pattern the chondrocranium (reviewed by Francis-West et al. 1998; Couly et al. 2002; Trainor et al. 2002). Furthermore, even between different arches and primordia, molecular studies have shown that the signalling interactions that control their development are unique (Ferguson et al. 2000). For example, Endothelin-1 differentially controls homeobox-containing gene expression in the branchial arches (see Sect. 5.2.2).

The first insight into tissue interactions that control patterning of the head was generated by chimeric grafting studies in which neural crest cells were grafted ectopically into another host. In amphibians this resulted in chimeric heads in which the neural crest cells formed structures specific to their origin rather than their host (Wagner 1949; Horstadius and Sellman 1946). (Also see Schneider and Helms, 2003 Science, 299: 565–568). Similar studies in the chick have been particularly informative. Transplantation of first arch neural crest (rhombomeres 1 and 2, mesencephalic) or more rostral mesencephalic neural crest into the caudal hindbrain yields ectopic proximal first arch structures, which are derived from the transplanted neural crest

(Noden 1983b). In contrast, first arch neural crest cells placed in a rostral position will integrate with the endogeneous neural crest populations to form a normal facial skeleton (Noden 1978a,b). Despite the latter evidence implicating environmental signals in the control of neural crest cell fate, at least in terms of the rostral populations, the results of these studies have been widely interpreted as indicating that all cranial neural crest cells are prepatterned as they emerge from the neural tube. This has lead to a succession of publications disputing this interpretation. For example, many studies have shown that neural crest cells can change their molecular fate when challenged with a new environment, arguing against the neural crest prepatterning model (reviewed by Francis-West et al. 1998). However, as reported in his original paper, Noden also noted the plasticity of first arch neural crest cells, which could become integrated into second and third arch structures. Other studies have shown that the neural crest cells are not important for the early establishment of the caudal arches, suggesting that environmental signals, as proposed by Noden for the upper face, control patterning (Noden 1983b, 1988; Veitch et al. 1999: Gavalas et al. 2001).

Further dissection of the tissue interactions controlling first arch patterning has shown that the developing brain is essential. Couly and colleagues found that the neural tube must be transplanted, in addition to the neural fold, which generates the neural crest. In the absence of the neural tube, the first arch crest just forms random cartilage structures or becomes integrated into the normal skeleton when transplanted caudally (Couly et al. 1998). This suggests that the neural tube either patterns the neural crest directly and/or modifies other environmental signals. More recently, the mid/hindbrain isthmus has been suggested to be the key signalling centre, again indicating that the neuroepithelium patterns the first arch neural crest precursors (Irving and Mason 1998; Trainor et al. 2002a). Grafts of the isthmus, but not the anterior hindbrain, in place of R4 induce the formation of duplicated proximal first arch structures, and this is associated with the down-regulation of the expression of *Hoxa2*, which controls patterning of the second arch, in the neural crest cells themselves. Although it is currently not totally clear whether the *Hoxa2* negative crest cells are induced in the surrounding host rhombomeres or arise within the original graft, the results suggest that neural crest cell identity may be specified by the isthmus. This would be consistent with other studies showing that the isthmus can pattern R1 (Irving and Mason 1998). Having acquired the appropriate molecular identity, the neural crest cells are patterned, or are "competent" to respond to environmental signals, to form the proximal first arch structures. This is consistent with Noden's original statement that "the neural crest precursors, while they are still part of the immature epithelium, acquire spatial patterning". The isthmus also controls patterning of the midbrain, and in both cases the signal can be mimicked, at least in part, by Fgf8 (Irving and Mason 1998; Trainor et al. 2002a).

A recent paper, which will undoubtly become a classic study, has shown that the endoderm is responsible for patterning the facial skeleton in the developing chick embryo. Endoderm ablations resulted in the loss of skeletal structures, whereas ectopic endoderm grafts in the developing face induced the formation of ectopic skeletal structures (Couly et al. 2002). Each region of endoderm was able to generate different skeletal structures, including the proximal first arch structures. This shows that differential expression of genes, which have not yet been identified, within the endoderm determine which skeletal structures will form. This study also shows that endodermal signals are needed to pattern the proximal first arch structures and that the isthmus, rather than directly determining patterning, is probably required to confer "competence" on the neural crest to respond to the endodermal signal. Indeed, the facial endoderm is only able to induce ectopic facial structures in the first branchial arch and more rostral regions and is unable to repattern more caudal arches, i.e. *Hox*-expressing regions, consistent with this idea. Also in keeping with this possibility is the inability of R4 neural crest (i.e. *Hoxa2*-expressing) to form proximal first arch structures in a first arch environment. Following rhombomere ablations, R4 neural crest cells subsequently migrate into and populate the first arch, maintaining their original Hox code; i.e. they express *Hoxa2*. In this case, neither first arch nor second arch structures develop (Couly et al. 1998). This shows that first arch endoderm cannot pattern second arch neural crest. As second arch structures do not form, it also suggests that unique second arch environmental signals are needed to generate second arch structures.

As mentioned above, Hoxa2 appears to be the major player in determining the second arch, which resembles the primitive ground state, versus the first arch structures. *Hoxa2* is expressed in the neural tube with a rostral limit of rhombomere 2. *Hoxa2* is also expressed in the resulting neural crest progeny with the exception of that arising from rhombomere 2 (Fig. 9; Prince and Lumsden 1994). Therefore, the second arch neural crest is distinguished from the first arch neural crest by the expression of *Hoxa2*. As discussed above, this expression appears to confer unique properties to the neural crest cells, making them insensitive to some environmental signals. Further evidence for this comes from ectopic transplantion studies. When the posterior hindbrain is transplanted rostrally into the posterior diencephalon, the ectopic neural crest cells maintain *Hoxa2* expression, giving rise to pericytes together with the appropriate neuronal and muscular connective tissues (Couly et al. 2002). However, they do not give rise to skeletal derivatives, showing that this crest population cannot respond appropriately (Couly et al. 2002).

The importance of Hoxa2 in branchial arch patterning has been shown by both loss- and gain-of-function studies. Hoxa2 is required and can be sufficient for second arch development. Thus, it has been proposed that loss-of-*Hoxa2* expression in the migrating neural crest cells may have changed the primitive ground state for lower jaw development. Loss of *Hoxa2* function in mice results in the transformation of second arch structures to those of the proximal first arch, i.e. the malleus, Meckel's cartilage, and tympanic ring (Gendron-Maguire et al. 1993; Rijli et al. 1993). This also results in the formation of a bone resembling the more primitive quadrate bone found in chicks in shape and position.

Conversely, misexpression of *Hoxa2* in the first arch results in the formation of second arch structures (Kanzler et al. 1998; Grammatopoulos et al. 2000; Pasqualetti et al. 2000). These latter studies have also shown that *Hoxa2* expression is required within the branchial arch itself and not only in the migratory and postmigratory neural crest cells (Grammatopoulos et al. 2000). This raises the question as to which environmental tissues control second arch patterning. The ectoderm is a candidate, as it expresses *Hoxa2* after the neural crest cells have migrated into the second branchial arch. Indeed, at one point it was proposed that patterning information was transferred from the ectomesenchyme to the overlying ectoderm (Hunt et al. 1991), although more recent studies by Couly et al. (1998) have shown that ectomesenchymal and ectodermal expression of *Hoxa2* do not always go hand in hand. These overexpression studies, like the original studies by Noden (1983b) have also revealed that as in the limb bud, skeletal structures pattern the developing musculature (Noden 1983b, 1986; Pasqueletti et al. 2000).

The *Hoxa2* gene inactivation or overexpression studies have also revealed the existence of a possible signalling centre within the developing face – the first branchial cleft. Following loss of *Hoxa2* function or overexpression of *Hoxa2*, the ectopic structures are mirror-image duplications and it has been suggested that there is an organising centre that controls the patterning of the crest populations (Rijli et al. 1993). This hypothesis seems totally feasible, and if true, the molecular player remains elusive. Candidate molecules include Shh, and possibly Fgf8, which are expressed in the first branchial cleft (Wall and Hogan 1995).

Mesodermal signals are important for maintenance, but not induction, of the appropriate Hox code, although the signalling factors involved remain unidentified (Trainor and Krumlauf 2000). Thus, individual second arch neural crest cells transplanted ectopically into a new environment do not retain their orginal Hox code but will do so if transplanted with mesoderm. This ability of neural crest cells to change molecular fate is also seen during neural tube regeneration and occurs naturally during development – for example in the few remaining neural crest cells from R3 and R5, which migrate into the adjacent pathways (Couly et al. 1996; Saldivar et al. 1997; Trainor and Krumlauf 2000). This plasticity probably also occurs in other neural crest cell populations where individual cells have meandered from the main stream, representing a refining process essential for the final sculpting of the arches. At first glance, the plasticity of neural crest seems in conflict with the studies by Couly et al. (1998) and others where the Hox code has been retained even in an ectopic environment (reviewed by Francis-West et al. 1998). However, these apparent discrepancies may reflect a community effect in which when large numbers of neural crest cells are present they will not alter their expression in an ectopic environment and can override new environmental signals. Indeed, neural crest cells do appear to communicate, as some of them contact each other during migration. Furthermore, they differentially express the signalling molecules, ephrins, which control cell sorting. Thus there appear to be competive interactions between the neural crest cells and the environment, with a fine balance determining the final outcome. When a neural crest cell is isolated it loses this communication, becoming susceptible to environmental signals.

Further evolution and modification of the jaw has required the acquisition of new genes or a change in the expression profile of existing genes, and the generation of knockouts has given us glimpses into how this may have occurred. For example, in the *RARαγ* double mouse mutants, the incus is linked to the alisphenoid resembling the more primitive bone, the epipterygoid, seen in reptiles (Lohnes et al. 1994). In addition, there is an ectopic bone in the upper jaw. This is reminiscent of the three pillars, pilae prooptica, metoptica, and antotica, which are found in reptiles and monotremes (lower mammals). Therefore a change in the response of cells to retinoic acid (RA) signalling may be responsible for some of the evolutionary transitions that have occurred in the head.

In addition, the combination of homeobox-containing genes, such as members of the Msx and Dlx families, has been proposed to control patterning of the face. Indeed, gene inactivation studies have confirmed their role during craniofacial development, with each gene controlling the development of a different region. In some cases, loss of gene function, as discussed for the RARs above, has resulted in the generation of ectopic structures resembling those seen in more primitive species, i.e. possibly atavistic changes whilst in other studies, specific structures have been lost (reviewed by Francis-West et al. 1998; Smith and Schneider 1998). Thus, it can be envisaged that the acquisition of a single gene, or more likely changes in its expression pattern, during evolution may allow additions/changes of skeletal structures in a "piece-meal" fashion. Therefore, the primordia and even the structures within the primordia can be seen as distinct "evolutionary" units. For example, inactivation of the ET-1 signalling pathway affects all the branchial neural crest-derived structures with the exception of the teeth (Kurihara et al. 1994).

Recent studies in the chick have suggested that the differential expression of Bmps and aldehyde 6 dehydrogenase, which generates RA, distinguishes the maxillary primordia from the frontonasal mass (Lee et al. 2001). *Bmp2* and *-4* are expressed in the maxillary primordia whilst *aldehyde 6 dehydrogenase* is expressed transiently in the presumptive frontonasal mass epithelium (Francis-West et al. 1994; Barlow and Francis-West 1997; Schneider et al. 2001). Loss of Bmp function with a simultaneous gain in RA signalling in the maxillary primordia results in the ectopic formation of frontonasal structures, except the nasal capsule, which is induced by the nasal placode, in place of the maxillary structures (Lee et al. 2001). Whether this is a true transformation, however, and whether the two distinct primordia are generated by the simple loss or addition of a signalling factor is yet to be conclusively proved. As discussed in Sect. 2.2, Bmps are present in the developing brain, where they may influence the frontonasal mass neural crest. In addition, earlier studies in which Bmp signalling alone had been blocked also resulted in an additional skeletal structure, together with an egg tooth, which is characteristic but not a definitive marker of the frontonasal mass (Bogardi 2002). The additional skeletal structure arose as a bifurcation of the endogeneous nasal septum, suggesting a very plausible explanation for this – that the frontonasal mass neural crest had mismigrated into the presumptive maxillary primordia.

5.2 Growth Factor Regulation of Outgrowth of the Facial Primordia

Initially the developing face consists of the mandibular primordia, which gives rise to the lower jaw, the maxillary, lateral, and medial (also known as frontonasal mass) processes which together form the upper jaw (Fig. 12). Co-ordinated outgrowth of the primordia and ultimately fusion are essential for normal development of the face, ensuring that it becomes a complete intact and functional unit. Defects in this process result in a number of facial abnormalities, which include clefting (see Sect. 5.3). Initially, the primordia are undifferentiated buds of mesenchyme consisting of the neural crest-derived mesenchyme, known as the ectomesenchyme, and mesoderm, covered by a layer of epithelium. Outgrowth of the facial primordia is controlled by epithelial-mesenchymal interactions such that removal of the epithelium results in truncation of facial development (Wedden 1987). The epithelium appears to be the passive recipient of instructions from the mesenchyme to control outgrowth. Hence, recombinations between the facial mesenchyme and ectopic epithelia, such as the limb bud ectoderm, all can support outgrowth and skeletal development (Richman and Tickle 1989, 1992). However, epithelial signals can be instructive, inducing odon-

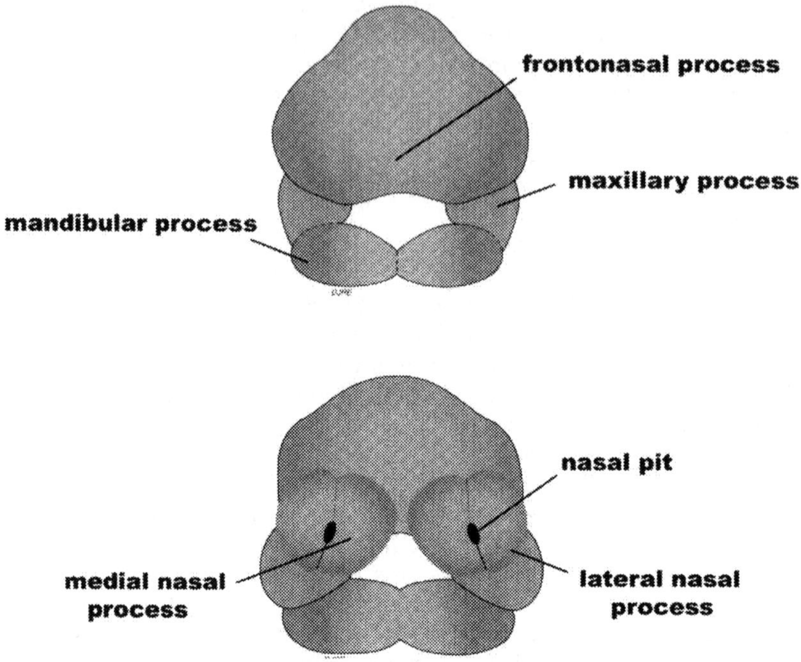

Fig. 12. Ilustration of the formation and subsequent early transformation of the facial processes. Initially the face is made up of five processes, the single frontonasal process, and the paired maxillary and mandibular processes. As development progresses the facial processes gradually become transformed, with the mandibular processes becoming merged in the midline. Following formation of the nasal placode and its subsequent invagination, the medial and lateral nasal processes are formed within the frontonasal processes. With continued transformation including the further growth and subsequent merging of processes, the face begins to take on a characteristic appearance. This includes a narrowing of the upper face such that the medial nasal processes fuse. Facial clefts can occur between the mandibular primordia, which result in cleft mandible. Clefts can also occur between the medial processes, which can result in the duplication of some nasal structures. The latter is often associated with neural defects such as exencephaly. Cleft lip can result from the failure of the lateral and medial nasal processes and maxillary primordia to fuse. See Sect. 5.3 and Table 3 for examples of factors that control these processes

togenesis in dental mesenchyme and skeletal formation such as the otic capsule (see Sect. 6.1; Mina and Kollar 1987; Lumsden 1988; reviewed by Hall and Miyake 1995, 2000).

Five key secreted growth factors or families of growth factors that control facial outgrowth, via regulation of cell proliferation and survival, have been identified. In addition, some of these factors have been implicated in patterning via their ability to regulate the expression of homeobox-containing genes. This group of factors includes endothelin-1, Shh, Bmps, Wnts, and Fgfs, and these are discussed below. In addition, retinoid signalling has been shown to be a crucial player in controlling frontonasal mass outgrowth.

The transcription factors that are mediators of growth factor signals include those of the homebox-containing gene family. The differential expression of the latter, which are related to the classical Hox gene family, have been proposed to control patterning of the odontogenic field and facial primordia. Until recently there was no conclusive evidence for a direct role in patterning, as loss-of-function experiments have not produced any transformation of skeletal elements to that of another region of the face, as occurs following *Hoxa2* gene inactivation. However a recent paper has shown that in the double *Dlx5/Dlx6* mouse mutant, the mandibular primordia is transformed to form the maxillary primordia (Depew et al. 2002. Science 298: 381–385). These transcription factors have also been implicated in a number of other biological processes such as epithelial-mesenchymal interactions (Msx1), cell migration (Pax6), cell survival (Cart1), and cell differentiation (Dlx5). The defects following gene inactivation can affect specific regions of the head such as occurs in the *Dlx2* mutant, or specific structures such as occurs in Riegers syndrome, which predominantly affects the teeth. Alternatively, the mutations can affect most of the head, as occurs in the *Msx1* mutant. These families and their effects on craniofacial development have been described elsewhere (Francis-West et al. 1998) and are summarised in Tables 1, 3, and 5. In addition, other transcription factors or factors belonging to transcription complexes such as AP-2 and p63 respectively, which control cell survival, have been shown to be crucial for facial development. The roles of these transcription factors are also summarised in Table 3.

5.2.1 Bone Morphogenetic Proteins

In addition to the role of this family in neural crest generation and apoptosis, Bmps also control patterning, outgrowth, and skeletal differentiation (see Sects. 4.1 and 4.2). Studies in chicks and mice have shown that the major players appear to be the epithelial signal, Bmp4, together with its closely related family member Bmp2, which is expressed in the underlying mesenchyme (Fig. 13; Francis-West et al. 1994; Bennett et al. 1995; Wall and Hogan 1995; Barlow and Francis-West 1997). These molecules control outgrowth and patterning of some regions of the facial primordia. Thus, application of Bmp2 or -4 to the chick mandibular or maxillary primordia induces the ectopic expression of *Msx1* and *-2*, and can result in the bifurcation of skeletal structures such as Meckel's cartilage derived from the mandibular primordia and the palatine bone formed in the maxillary primordia (Barlow and Francis-West 1997). The bifurcations are associated with an increase in cell proliferation and expansion of the primordia. Conversely, loss of Bmp signalling, by application of the Bmp antagonist, noggin, truncates the maxillary primordium, and results in loss of *Msx1* gene expression together with the loss of skeletal structures. Gene inactivation of *Msx1* in mice also affects the development of the maxillary bones, including the loss of palatine and maxillary shelves, suggesting that during facial development, *Msx1* is a key downstream mediator of Bmp signalling, as occurs in other developmental systems (Fig. 13; Satokata and Maas 1994; Monsoro-Burq et al. 1996; Houzelstein et al. 1997; Bei and Maas 1998; Tucker et al. 1998a; Yamamoto et al. 2000). Homozygous *Bmp4* mutant mice die before E9.5, too soon for the role of Bmp4 during facial outgrowth in mice to be evaluated (Winnier et al. 1995; Lawson et al. 1999). However, consistent with the proposed role in outgrowth, which has been shown in chicks, heterozygote *Bmp4* mutant mice have slightly truncated snouts (Dunn et al. 1997). Seemingly, in apparent conflict to this, loss of function of the Bmp antagonists, *chordin* and *noggin*, which in general are expressed in domains in the epithelium complementary to that of *Bmp4*, also reduces mandible outgrowth (Stottmann et al. 2001). This is in part due to loss of *Fgf8*, which controls outgrowth. The decreased outgrowth may also be related to high levels of Bmp signalling, which in the facial primordia and other systems has been shown to induce apoptosis (Barlow and Francis-West 1997). Thus, an increase in Bmp activity may directly induce apoptosis of the mesenchymal cells via an Fgf8-independent route (Stottmann et al. 2001).

In addition to regulating facial outgrowth, Bmp4 has also been shown to control the position and patterning of the odontogenic field through antagonistic interactions with Fgf8, suggesting it has a direct role in patterning (Fig. 13; Neubüser et al. 1997; Tucker et al. 1998b). In addition, Bmp4 is also thought to specify the division between the maxillary and lateral nasal process during very early facial development (Shigetani et al. 2000).

Bmp5 and -7, which belong to a different subgroup of the Bmp family, also control craniofacial growth (Dudley et al. 1995; Solloway and Robertson 1999). Minor craniofacial defects occur as a result of gene inactivation of *Bmp7* or in the natural mouse *Bmp5 short-eared* mutant, where defects are predominantly localised to the eyes and ears respectively (Kingsley et al. 1992; Dudley et al. 1995; Luo et al. 1995). However, loss of both *Bmp5* and *Bmp7* function results in more severe and widespread craniofacial abnormalities, affecting the neural tube, which is slower to close, and the branchial arches, which are smaller (Solloway and Robertson 1999). Both *Bmp5* and *Bmp7* are initially expressed in the rostral neuroectoderm, mesoderm, and in the ectoderm. Later, they are expressed in the ectoderm of the branchial arches, with *Bmp5* being confined to the distal tips (Solloway and Robertson 1999; Wall and Hogan 1995). The reduced size of the branchial arches appears to be due, at least in part, to increased cell death in the ectoderm. However, the expression of *Fgf8*, a factor that controls outgrowth and cell survival, appears to be unaffected (Solloway and Robertson 1999).

5.2.2 Endothelins

One member of this family, endothelin-1 (ET-1) has been clearly demonstrated to be essential for lower jaw development in both teleosts and amniotes, and hence its role has been conserved for over 350 million years. This necessity may reflect the small number of endothelin family members with very distinct and complementary developmental expression patterns. It has even been proposed that gain of ET-1 signalling in the anterior arches was a criticial step for the evolution of jawed vertebrates (Miller et al. 2000). ET-1 is a small peptide consisting of 21 amino acids and belongs to a family comprising three members, ET-1, -2, and -3. The endothelins are synthesized as much larger precursors, which are then proteolytically activated in two steps. First, they are cleaved into an inactive precursor, called a big endothelin, by a furin-like processing protease. They are subsequently cleaved at a conserved Trp21 residue by either endothelin converting enzyme-1 (ECE-1) or -2, which belong to the type II membrane-bound metalloprotease family, to generate the biologically active molecule. They signal through one of two G-protein coupled receptors, ET_A and ET_B, which activate the phospholipase C/Ca^{2+} pathway. ET-1 and -2 can signal through both receptors, whilst ET-3 specifically signals via ET_B. Gene inactivation studies in mice

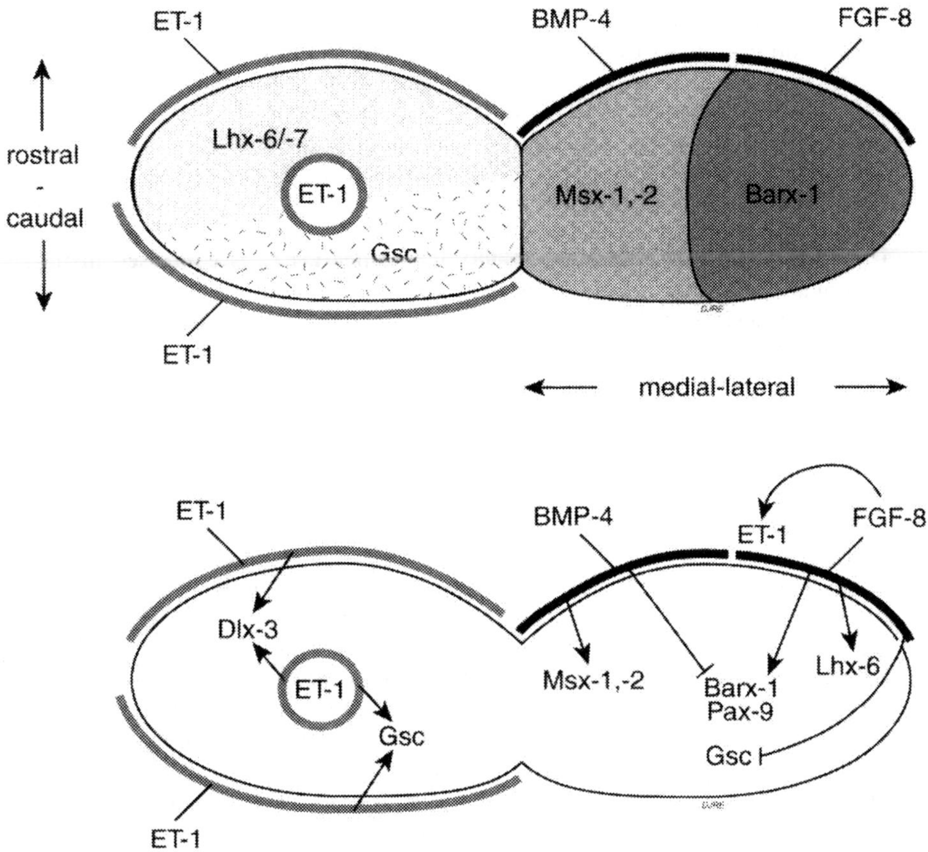

Fig. 13. Schematic summarising gene expression patterns (*upper panel*) and signalling interactions (*lower*) during the development of the mandibular primordia. The *thick bars* represent the epithelium; the *circle* in the centre of the mandibular primordium on the left-hand side represents the mesoderm. See Sect. 5.2 and Table 3 for more details

have shown that this family of peptides controls crucial aspects of neural crest development. ET-1 and ET_A are required for normal craniofacial development and development of the aortic arches and heart, whilst ET3 and ET_B are required for development of the trunk and vagal neural crest lineages. Thus, loss of *ET-3* and ET_B function in mice results in a spotted coat phenotype due to the lack of pigmentation, and a lack of hindgut ganglia (Baynash et al. 1994; Hosoda et al. 1994). Mutations in *ET-3* and ET_B have also been identified in humans, in familial Hirschsprungs syndrome, which has a similar phenotype to that seen in the mutant mice (Puffenberger et al. 1994; Edery et al. 1996; Hofstra et al. 1996). In contrast, gene inactivation of *ET-1* and ET_A in mice affects cranial and cardiac development, whilst loss of ECE-1 encompasses the phenotypes seen following loss of both *ET-1* and *ET-3* function, suggesting that ECE-1 is the major regulator of both signalling pathways (Yanagisawa et al. 1998b).

ET-1 is expressed in the branchial arch epithelium including the endoderm of the pharyngeal pouches, endocardium of heart chambers, the paraxial mesoderm, and some of its derivatives such as the arch vessel endothelium and menigeal vessels around the brain in chicks, mice, humans, and zebrafish (Fig. 13; Kurihara et al. 1994, 1995; Barni et al. 1995; Chan et al. 1995; Maemura et al. 1996; Brand et al. 1998; Clouthier et al. 1998; Kempf et al. 1998; Nataf et al. 1998; Miller et al. 2000). In general, the receptor, ET_A, is expressed in a reciprocal fashion. Thus, in mice and humans but not chicks, ET_A is expressed in the rhombomeric migrating crest whereas later in mice, chicks, and humans, ET_A mRNA transcripts are found in the ectomesenchymal derivatives of this neural crest such as the vascular smooth muscle cells (Brand et al. 1998; Clouthier et al. 1998; Nataf et al.1998). In the heart, ET_A is expressed in the myocardium and mesenchyme of the aortic arches (Clouthier et al. 1998; Kempf et al. 1998). *ECE-1* transcripts are found in the head mesenchyme, heart, and the branchial arch surface epithelium (Yanagisawa et al. 1998b).

Mutational inactivation of *ET-1*, ET_A, or *ECE*-1 results in very similar craniofacial and heart phenotypes (Kurihara et al.1994, 1995; Clouthier et al. 1998a,b; Yanagisawa et al. 1998). Overall there is a gross defect in epithelial-ectomesenchymal interactions. The craniofacial phenotype is marked by a severe aplasia of the first and second branchial arches, which is clearly apparent at E10.5. This results in the loss/reduction of Meckel's cartilage and mandible bones together with defects in maxillary derivatives, resulting in cleft palate. Second arch derivatives such as the hyoid bone, which forms Reichert's cartilage in the adult, are malformed (Kurihara et al. 1994). The outer and middle ear are absent or hypoplastic. In contrast to these skeletal defects, the teeth are intact, consistent with the proposal that the dentition and craniofacial skeletal derivatives are distinct evolutionary units which can be uncoupled (Smith and Hall 1993; Kurihara et al. 1994; Yanagisawa et al. 1998b). The zebrafish mutant *sucker* (*suc*) results from a missense mutation in *ET-1*, which changes the charged residue asparagine 8 to the non-charged valine. This mutation appears to be a loss of function, as biochemical analyses have shown that it is one of five amino acids known to be critical for ET-1 signalling (Tam et al. 1994). Furthermore, recombinant ET-1 can rescue the defects in the *suc* mutant. As in mice mutants, there are lower jaw abnormalities (Miller et al. 2000).

There are also heart defects in mice, which resemble those seen after loss of neural crest – abnormal remodelling of the arteries, interruption of the aorta, and persistent truncus arteriosus (Kurihara et al. 1994; Yanagisawa et al. 1998a). Similar craniofacial and heart phenotypes have also been obtained in the developing chick embryo following administration of the ET-1 antagonist RU69986 (Kempf et al. 1998). Again the lower jaw and second arch are smaller, and development of the thymus and arteries is affected – they are either absent or the arteries may be abnormally bifurcated (Kempf et al. 1998). In contrast, circulatory defects have not been observed in the zebrafish *suc* mutant (Miller et al. 2000).

In addition to skeletal defects, the sublingual and submandibular ducts, thyroid, parathyroids, and thymus, which are dependent on epithelial-mesenchymal interactions for their development, are absent or hypoplastic (Kurihara et al. 1994, 1995). The thymus also does not descend. In zebrafish, some of these defects have been preliminarily linked to a smaller domain of *Shh* and the failure to form the midline of the diverticulum (Miller et al. 2000). The tongue is also partially missing whilst the facial muscles are disorganised and hypoplastic. The latter defect is probably a secondary consequence of the neural crest and/or skeletal defects, as the ET_A receptor is not expressed in presumptive myogenic cells. Studies in zebrafish have shown that there are differential defects in muscle development in the different arches. ET-1 signalling is required for activation of the expression of *MyoD*, a myogenic determination factor, in the presumptive posterior pharyngeal arch muscles but is not required in the second and first arch muscles – these muscles develop but are hypoplastic.

Analysis of mice and zebrafish mutants has shown that ET-1 signalling is not needed for neural crest migration but is essential for cell survival and proliferation, at least within the first arch (Clouthier et al. 1998, 2000; Miller et al. 2000). ET-1 differentially controls the expression of homeobox genes in the different arches, with defects occuring primarily in the distal regions where *ET-1* expression is highest (Clouthier et al. 2000). In the mouse, ET-1 controls the expression of *Dlx3*, *Dlx6*, and *Goosecoid* (*Gsc*), together with the basic-loop-helix proteins, *dHAND* (also known as hed/thing 1) and *eHAND* (also known as hxt/thing 1) in the first and second arches (Fig. 13; Clouthier et al. 1998, 2000; Thomas et al. 1998; Charité et al. 2001). The loss of *dHAND* expression is secondary to the loss of *Dlx6*, which induces *dHAND* expression (Charité et al. 2001). Other genes are differentially regulated within the two arches – the expression of *Barx1*, ET_A, and *Dlx2* is unaffected in the first arch but is lost in the second arch (Clouthier et al. 2000). All these genes, with the exception of *Dlx2*, are expressed in post-migratory neural crest, suggesting that they are switched on in response to ET-1 signalling from the epithelium. In contrast, the expression of *Mhox*, *Dlx1*, *Msx1*, and *Ufd1* are unaffected, at least at the transcriptional level (Clouthier et al. 1998, 2000; Thomas et al. 1998). The latter Ufd1 degrades ubiquitinated proteins, and its deletion at one time was thought to be a causative factor in **CATCH22** (cardiac anomaly, abnormal face, thymic hypoplasia, cleft palate, hypocalcemia and chromosome 22 deletions) syndromes, although recent data points to other genes being the major and more probable players in this syndrome (see Sect. 8.3). Analysis of the *suc* mutant has revealed a similar cascade of regulatory loops with the expression of *dHAND*, *MsxE*, *Gsc*, and *Dlx3* all being dependent on ET-1 function. In zebrafish, ET-1 signalling has also been shown to regulate the expression of *EphA3*, the ephrin receptor tyrosine kinase, and *Dlx2* in the first arch, in addition to the second arch regulation of *Dlx2*, which is observed in mice.

Consistent with the above, the phenotypes of the *Gsc*, *Dlx2*, *dHAND* mutants overlap with the *ET-1* knockout. For example, *Dlx2* mouse mutants have defects in the second arch derivatives (Qiu et al. 1995, 1997). The phenotypic overlap between the *Gsc* and *ET-1* mutants is restricted to the branchial arches such as the bones of the middle ear (Rivera-Pérez et al. 1995, 1999; Yamada et al. 1995). Therefore, the defects in the nasal structures that occur in the *Gsc* mutant are not seen in the ET-1 knockout mouse. This is consistent with the restriction of *ET-1* expression to the branchial arches and further supports the concept that each region of the head can be viewed as a distinct module with its own unique set of signalling interactions. Supportive of this, RA treatment of E8 mice embryos down-regulates *Gsc* expression in the branchial arches but not in the nasal mesenchyme, either reflecting the differential regulation of *Gsc* itself in the different regions or the differential effects of RA treatment on different populations of neural crest (Zhu et al. 1997). The *dHAND* and *ET-1* mutants also show striking parallels. In the *dHAND* null mutants, there is increased apoptosis; the first and second arches are hypoplastic whilst the third and fourth arches do not form (Thomas et al. 1998). Proliferation is unaffected, at least at E9.5. Also, similarly to the *ET-1* mutant, the expression of *Mox1* and *Dlx2* is unaffected in the mandibular arch of the *dHAND* mutant mouse. However, paradoxically whilst the expression of *Msx1* is unchanged in the *ET-1* mutant, it is down-regulated in *dHAND* mutants. This apparent discrepancy may be related to the residual *dHAND* expression in the midline of the branchial arches following loss of function of ET-1 signalling – this is likely to be sufficient to induce/maintain *Msx1* expression. *ET-1* expression is unaffected in *dHAND* mutants, showing that there is not an auto-regulatory loop between *dHAND* and *ET-1* expression (Thomas et al. 1998). *EHAND*, which is down-regulated by loss of ET-1 function, is unaffected in *dHAND* mutants, suggesting that there are at least two distinct pathways activated by ET-1 signalling.

The *Dlx3* homozygous mutant dies between E9.5 and E10.5 due to placental defects, and therefore it is unclear how much loss of *Dlx3* function contributes to the ET-1 phenotypes (Morasso et al. 1999). Mutation of the *DLX3* gene in humans can result in Tricho-dento-osseous syndrome, characterised by taurodontism (elongation of the pulp chambers) and enamel hypoplasia, with resultant abcess formation as the chambers become exposed, together with curly hair and bone defects. In the latter, the cranial bones become thicker, ultimately resulting in the loss of the sinuses (Price et al. 1998). It is unclear whether the bone or dental defects are related to loss of ET-1 signalling as they are relatively late onset defects, which would not be picked up in the *ET-1* mutant mouse.

5.2.3 Fibroblast Growth Factors

A number of members of the Fgf family, which consists of over 22 members, are expressed in the developing face (reviewed by Francis-West et al. 1998; see also Bachler and Neubüser 2001; Munoz-Sanjuan et al. 2001). These factors are expressed either in the epithelium, as is the case with *Fgf8* and *Fgf2*, or alternatively are localised to the mesenchyme as is the case with *Fgf5* and *cFHF4*. The key member is *Fgf8*, which is expressed in restricted regions of the epithelium (Fig. 13). The function of Fgf8 appears to be controlling patterning via regulation of homeobox-gene expression and epithelial-mesenchymal interactions, i.e. outgrowth. Recent work has also implicated Fgf8 signalling defects in DiGeorge syndrome as loss of function of *Fgf8* affects both the development of the branchial arches and cardiac structures (Abu-Issa et al. 2002. Development 129: 4613–4625; Frank et al. 2002. Development 129: 4591–4603). Also see Trokovic et al. 2003. Genes and Dev. 117: 141–153.

The role of Fgf signalling in outgrowth has been clearly demonstrated by conditional inactivation of *Fgf8* function and overexpression of the dominant-negative receptor Fgfr2. *Fgf8* is expressed in the presumptive first arch ectoderm (Trumpp et al. 1999; Shigetani et al. 2000). As the arch develops, *Fgf8* expression becomes restricted to lateral and caudal regions of the mandibular and maxillary primordia, respectively (Fig. 13; Helms et al. 1997; Richman et al. 1997; Francis-West et al. 1998; Barlow et al. 1999; Shigetani et al. 2000). Inactivation of *Fgf8* expression, prior to E9 as the arches are forming, results in the loss of the mandibular and maxillary derivatives except the most distal structures, the symphyseal region of Meckel's cartilage and the maxilla, respectively (Trumpp et al. 1999). There is also microglossia, disorganisation of the tongue. The skeletal losses are associated with smaller arch development, at least in part due to increased apoptosis (Trumpp et al. 1999). Analysis of gene expression has shown that patterning within the arch is relatively normal. Thus *Dlx1*, *-2*, *-5*, *Bmp4*, *Msx1/2*, *dHAND*, *eHAND*, *Lhx7*, and *Pitx1* are all expressed in their normal domains.

In contrast, with the exception of the proximal domain where *Fgf8* is transiently expressed, *Lhx6*, *ET-1*, and *Barx1* transcripts are absent, showing that Fgf8 signalling is needed to induce their expression (Fig. 13). The maintenance of their expression in the proximal region also shows that Fgf8 function is required to activate but not maintain their expression. This role may be taken over by ET-1, which has been shown to control *Gsc* expression.

Fgf8 can be expressed as one of seven isoforms. Of these, Fgf8b has been clearly identified in the branchial arch and nasal pit ectoderm, where it has been proposed to signal and activate Fgfr2c in the underlying mesenchyme (MacArthur et al. 1995). Transgenic studies in which a dominant-negative Fgfr2 soluble receptor has been misexpressed in mice have also implicated this receptor as being important for facial outgrowth. The resulting mice have smaller jaws and cleft palate (Celli et al. 1998). In contrast, the Fgfr2 (IIIb) receptor isoform is expressed in the epithelium, where it may respond to Fgf1, -3, -7 or -10 signalling from the ectoderm and/or underlying mesenchyme. Loss of *Fgfr2 (IIIb)* function in mice results in a thinner mandible, cleft palate, and inner ear defects, with the consequent abnormalities in the otic capsule, which is induced by the inner ear (De Moerlooze et al. 2000). In addition there are abnormalities in tooth development, which is arrested at the bud stage; Rathke's pouch is absent due to apoptosis of the pouch epithelium, and there are salivary gland defects. There is also premature fusion of the suture between the parietal and temporal bones. Fgf3 is the candidate factor to control inner ear development through Fgfr2 (IIIb) as it is expressed in the otic ectoderm and later in the inner ear (Mansour et al. 1993; Mahmood et al. 1996). Furthermore, gene inactivation of Fgf3 also results in inner ear defects (Mansour et al. 1993). Fgf7 is expressed in the salivary gland mesenchyme and hence may be the key factor controlling salivary gland development (Mason et al. 1994).

Fgf signalling is in part mediated by Shp2, a tyrosine kinase, which is recruited to the docking protein Frs2 following phosphorylation of Frs2 by Fgf receptors. Chimeric mice made between Shp2-deficient and wild-type cells have a similar phenotype to Fgfr1 chimeras (Saxton et al. 2000). In the branchial arches, Shp2-deficient and wild-type cells segregate, suggesting that Shp2 may control cell adhesion and/or migration.

5.2.4 Sonic Hedgehog

Sonic Hedgehog (Shh) is crucial for development and outgrowth of the fontonasal mass as demonstrated by the *Shh* knockout mouse phenotype, in which the upper face is replaced by a proboscis-like extension lacking any skeletal structures (Chiang et al. 1996). In addition, there is cyclopia which as discussed earlier (Sect. 2.2) is due to the requirement of Shh signalling to pattern the midline of the neuroepithelium. In contrast, the branchial arches are relatively normal in the *Shh* knockout.

RA signalling is initially essential to maintain *Shh*, together with *Fgf8*, expression in the developing forebrain and overlying epithelium. RA is produced transiently in the presumptive frontonasal mass epithelium by aldehyde dehydrogenase 6 between stages 10 and 12 in the chick (Hamburger and Hamilton 1952). Application of retinoid antagonists at stage 10 in the chick results in a down-regulation of *Fgf8* and *Shh* expression in the forebrain and epithelium (Schneider et al. 2001). This is followed by a decrease in cell proliferation and increased cell death, eventually leading to total truncation of frontonasal mass development and hypotelorism. This truncation can be rescued by co-application of Shh and Fgf8, suggesting that these are downstream mediators of retinoid signalling. This retinoid signalling cascade is mediated by the migrating neural crest cells, which express the retinoic acid receptors, RARα, RXRβ and RXRγ (Rowe and Brickell 1995; Rowe et al. 1991, 1992; Hoover and Glover 1998). Likewise, loss of RARα function results in increased cell death and hypoplasia of the frontonasal mass in mice (Lohnes et al. 1994).

Later in development, *Shh* is expressed in the epithelium of the frontonasal mass, maxillary primordia, and oral cavity where it controls outgrowth (Barlow and Francis-West 1997; Helms et al. 1997; Hu and Helms 1999). Application of Shh blocking antibodies reduces outgrowth of the frontonasal mass, resulting in hypotelorism with clefting between the frontonasal mass and maxillary primordia. In contrast, gain of function by application of recombinant Shh protein or by retroviral misexpression of Shh increases proliferation in the region of induced *Bmp2* expression, and expands the upper face, resulting in either hypertelorism or bifurcation of skeletal structures (Hu and Helms 1999). The expansion of the frontonasal mass is secondarily associated with cleft palate as the palatal shelves fail to meet. In the branchial arches, Shh regulates the expression of *Tbx1*, a gene implicated in DiGeorge syndrome (see Sect. 8.3; Garg et al. 2001) whilst the expression of *Shh* in some regions of the mandibular primordia is regulated by the expression of *Prx1* and *-2* in the underlying mesenchyme (ten Berge et al. 2001). As in the upper face, Shh is essential for outgrowth, and thus blocking Shh function using vertrum alkaloids or by misexpression of Ptc reduces outgrowth of the lower jaw (ten Berge et al. 2001).

5.2.5
Wnts

In contrast to the key role played by members of the Wnt family in patterning other developmental systems, such as the limb and brain, this gene family has not been shown to pattern the skeletal structures of the facial primordia, but one member, Wnt5a, is essential for facial outgrowth. *Wnt5a* is expressed in the facial mesenchyme, with higher levels distally (Yamaguchi et al. 1999). Loss of function in mice by gene inactivation results in truncation of the jaw, in addition to truncation of the limb and axial structures. Hence the last structures to form, such as the digits and tail, are either not present or are much smaller. The other skeletal structures are also much smaller. In the paraxial and limb mesoderm, this has been shown to be due to decreased proliferation, and presumably, similar mechanisms operate in the face. Another member of the family, Wnt7b, has been shown to repress *Shh* expression in the mandibular epithelium defining the boundary between oral and odontogenic ectoderm in mice (Sarkar et al. 2000).

5.2.6
Interplay of Growth Factor Signalling

The studies to date have suggested that the growth factors co-ordinately regulate growth and patterning of the developing face. As in other developmental systems there is an interdependence of growth factor signalling and expression (Fig. 13). For example, the expression of *ET-1* is dependent on Fgf8 function, at least in some regions of the developing face. In contrast, antagonistic signalling interactions between Fgf8 and Bmp4 demarcate the boundaries of gene expression for a number of homeobox-containing genes, determining the position of the odontogenic field and odontogenic patterning (Neubüser et al. 1997; Tucker et al. 1998b). This was shown by *in vitro* studies where, in the absence of the epithelium, Bmp2 blocks the ability of Fgf8 to induce/maintain the expression of Fgf8-responsive genes such as *Barx1*, *Pax9*, and *Dlx5* (Fig. 13; Neubüser et al. 1997; Tucker et al. 1998b; Barlow et al. 1999; Bogardi et al. 2002). Thus, the differential distribution of *Fgf8* and *Bmp4* is associated with the differential expression of homeobox-containing genes in the ectoderm and mesenchyme. In fact, the same factor can affect the expression of homeobox genes in the ectoderm and mesenchyme in opposing ways. For example, *Fgf8* overlies and regulates *Dlx2* and *Barx1* expression in the mesenchyme whilst *Bmp4* is co-expressed with *Dlx2* and *Barx1* in the epithelium (Barlow et al. 1999: Thomas et al. 2000). Experiments in the mouse mandibular primordia have shown that Bmp4 induces, whilst Fgf8 represses *Dlx2* expression in the epithelium, in contrast to their effects in the mesenchyme (Thomas et al. 2000). Bmp4 signalling controls epithelial *Dlx2* expression via a planar mechanism, whereas Fgf8 represses *Dlx2* expression in the epithelium via a signalling relay through the mesenchyme (Thomas et al. 2000).

Studies in the chick maxillary primordia have shown that Bmp4 not only antagonises Fgf8 signalling but that Bmp4 controls *Fgf8* expression in a stage-dependent fashion. Shigetani et al. (2000) showed that *Fgf8* is expressed in the presumptive mandibular-maxillary ectoderm at stage 10 prior to the arrival of the neural crest. *Bmp4* is expressed slightly later at stage 12 in an adjacent domain. At this very early stage of facial development, the onset of Bmp signalling acts to restrict the domain of *Fgf8* expression (Shigetani et al. 2000). A similar effect is obtained by an increase in Bmp activity in the mandibular primordia following inactivation of the Bmp antagonists, *chordin* and *noggin*, in the mandibular primordia (Stottmann et al. 2001). In contrast, at stage 15, in the chick maxillary primordia, Bmp signalling is essential for the expression of *Fgf8* such that loss of Bmp signalling results in a smaller domain or loss of *Fgf8* expression (Bogardi et al. 2002). Later, by stage 20, there is yet another switch with loss of Bmp function resulting in the expansion of *Fgf8* expression (Bogardi et al. 2002). Therefore, at this stage Bmp signalling again represses the expression of *Fgf8*. The expansion of *Fgf8* expression has analogies to the limb bud where repression of Bmp signalling, again through misexpression of Bmp antagonists, increases the length of the apical ectodermal ridge, which expresses *Fgf8* and *Fgf4* (Capdevila et al. 1999; Merino et al. 1999; Pizette and Niswander 1999; Zuniga et al. 1999). By stage 24, *Fgf8* expression is independent of Bmp signalling in the maxillary primordia. In contrast to Bmp4, Shh does not appear to play a role in the regulation of *Fgf8* expression, at least between E9.5 and 10.5 in the mouse mandibular primordia (ten Berge et al. 2001).

The expansion of *Fgf8* expression in the maxillary primordia following loss of Bmp function is associated with an extension of the expression of *Barx1* (Bogardi et al. 2002). This has also been observed in the mouse mandibular primordia where the extension of *Barx1* expression is associated with transformation of tooth identity from incisors to molars (Tucker et al. 1998b). However, the study in the chick maxillary primordia has shown that the equivalent expansion does not result in the transformation of skeletal structures (Bogardi et al. 2002).

This Bmp antagonism of Fgf signalling is not unique to the developing facial primordia, having also been observed during feather, limb, and lung development (Jung et al.1998; Niswander and Martin 1993; Noramly and Morgan 1998; Weaver et al. 2000). Analysis of feather bud development has also shown that Bmp signalling can regulate Fgf function in additional ways via the regulation of the expression of Fgf receptors (Normaly and Morgan 1998). Whether similar mechanisms of antagonism also occur during face development are at present unclear. Interestingly, during lung development, Fgf10 induces the expression of *Bmp4* in the endoderm closest to the Fgf10 source, i.e. at the highest concentrations of Fgf10 (Weaver et al. 2000). This is another example of the induction of an antagonist, which will limit the signalling activity of a growth factor as seen in many developmental systems, usually by the induction of the expression of a secreted antagonist rather than another growth factor. As *Bmp4* is expressed after *Fgf8*, and in an adjacent domain, during mandibular and maxillary development, this raises the possibility that Fgf8 may also induce *Bmp4* expression (Shigetani et al. 2000). Pitx2 is also important to maintain the differential expression of *Fgf8* and *Bmp4*. Gene inactivation of *Pitx2* in mice results in the loss of *Fgf8* expression and expansion of *Bmp4*; the latter is associated with an expansion of *Msx1* and *Msx2* expression in the underlying mesenchyme (Lu et al. 1999b).

In addition to the antagonistic interactions between the growth factor families, their function is also modulated by the expression of secreted antagonists. For example, by stage 22, the Bmp antagonist *gremlin* is expressed in the posterior mesenchyme of the maxillary primordia, which will limit Bmp activity (Bardot et al. 2001). The other Bmp antagonists, *chordin* and *noggin*, are expressed in the mandibular primordia, where their expression is induced by Bmp signalling (Stottmann et al. 2001). In a similar fashion the Fgf antagonists, *sprouty-2* and *-4*, are expressed in the facial mesenchyme, where their expression may be regulated by the expression of Fgfs themselves (de Maximy et al. 1999; Chambers and Mason 2000).

As Bmp4 and Fgf8 appear to be major signalling factors differentially controlling homeobox gene expression, one question is that of how their complementary expression domains arise. *Fgf8* is expressed at stage 10, prior to neural crest arrival whilst *Bmp4* is expressed slightly later (Shigetani et al. 2000). As the maxillary and mandibular primordia are composed of two distinct populations of neural crest – from the hindbrain and the midbrain – the complementary expression domains of *Fgf8* and *Bmp4* may be related to the distinct combinations of neural crest, which induces or maintains their expression in the overlying epithelium (Le Lièvre and Le Douarin 1975; Lumsden et al. 1991; Scherson et al. 1993; Couly et al. 1996; Köntges and Lumsden 1996). Alternatively, the expression domains of *Fgf8* and *Bmp4* expression may be totally intrinsic to the epithelium. This would then determine homeobox gene expression in the underlying ectomesenchyme regardless of the different origins of the neural crest cells.

Whilst Bmp4-Fgf8 interactions control the medial-lateral axis of the mandibular primordia, Fgf8 and ET-1 signalling interactions have been proposed to control rostral-caudal polarity (Fig. 13; Tucker et al. 1999). The rostral and caudal axes are marked by the expression of *Lhx7/-8* and *Gsc*, respectively. Fgf8 signalling induces *Lhx6/7* expression whilst *Gsc* expression is activated/maintained by ET-1. In addition, *Gsc* expression is repressed by Fgf8 signalling which defines and localises *Gsc* expression to the caudal region of the arch in a complementary domain to *Lhx7* (Fig. 13).

5.3
Fusion of the Facial Primordia

The facial primordia arise as relatively discrete buds of tissue, which must fuse to generate the intact face (Fig. 12). The merging of the primordia is associated with epithelial cell death, migration, and/or epithelial-mesenchymal transformation with the distinct primordia fusing in a variety of ways. For example, fusion of the secondary palate involves all three mechanisms whilst fusion of the paired mandibular primordia predominantly involves epithelial migration (Pratt and Martin 1975; Kosaka and Eto 1986; Fitchett and Hay 1989; Carette and Ferguson 1992; Griffith and Hay 1992; Shuler 1995; Chai et al. 1997; Martínez-Álvarez et al. 2000a,b). Clefting is the failure of the distinct primordia to fuse and can occur between any of the primordia. During development of the face, fusion occurs between the paired mandibular primordia which generates a united lower jaw, the paired medial nasal processes fuse to narrow the upper jaw, whilst fusion between the maxillary and mandibular primordia determines the size of the mouth. Macrostomia, an enlarged mouth, results from too little fusion whereas microstomia is caused by too much fusion. The lateral nasal process and the medial nasal process merge together with the maxillary primordia to generate the primary palate. Defects in this process result in cleft lip, a relatively common craniofacial abnormality affecting 1 in 950 births. Finally, the paired maxillary primordia must fuse to separate the oral and nasal cavities. Failure of the latter results in cleft palate, which occurs in 1/700 to 1/1,000 births.

Clefting can be a result of a defect in neural crest generation, migration, survival, and/or proliferation. These defects would result in the formation of a smaller primordium which would fail to grow to the appropriate size and hence fuse. In addition, the epithelial seam may fail to migrate and/or break down, preventing the formation of a united structure. Clefting is multifactorial, being associated with teratogenic insults, genetic factors, and human syndromes (reviewed by Schutte and Murray 1999). For example, males and females have different susceptibilities to cleft lip and palate. The former occurs more frequently in males whilst cleft palate occurs more often in females. The latter is associated with delayed elevation of palatal shelves in females, which occurs during the eighth week as opposed to the seventh week in males (Burdi and Silvey 1969a,b). The face will have widened during this period and thus, it is more probable that the palatal shelves will not make contact.

Median clefts of the face are extremely rare and arise due to failure of the two medial processes (or frontonasal process) to merge. Grafting experiments in the chick have shown that initially the frontonasal mass has the potential to generate two or three nasal septums when divided into thirds (Wedden and Tickle 1986). Thus, there appears to be a general "ground plan" of all of the frontonasal mesenchyme to generate the nasal septum. During normal development this potential is inhibited in the lateral mesenchyme, presumably due to a lateral inhibition mechanism from the midline. In $Alx4^{-/-}/Cart1^{+/-}$ mutant mice, the facial cleft with two nasal septa can clearly be observed (Qu et al. 1999). A midline cleft also occurs in the *RARαγ* double mutants and is associated with reduced outgrowth and development of the skeletal structures (Lohnes et al. 1994). In the *platelet-derived growth factor receptor*, (*PDGFRα*) mouse knockout or the *Patch* (*Ph*) mutant which has a deletion in PDGFRα, the midline defect has been associated with abnormal neural crest migration and increased apoptosis of the post-migratory neural crest cells (Morrison-Graham et al. 1992; Schatteman et al. 1992; Soriano 1997; Robbins et al. 1999; Sun et al. 2000). In these mutants the mandibular primordia also fail to fuse, and in the *Patch* mutant this has been correlated with a down-regulation in the levels of the active metalloproteinase MMP-2, and of its activating enzyme, MT-MMP, which are involved in matrix remodelling (Robbins et al. 1999; Sun et al. 2000). *MMP-2* is normally expressed at fusion points in the facial primordia where it may play role in fusion, although it is not essential as shown by the lack of facial phenotype in the *MMP-2* mouse knockouts (Iamaroon et al. 1996; Itoh et al. 1997).

The secondary palate arises from shelves of the maxillary primordia, which initially are found on the lateral side of the tongue, and in mammals point downwards into the oral cavity (Fig. 14). In the developing mouse, the secondary palatal shelves are apparent at E12. As the mandibular primordia grows, this lowers the tongue, leaving room for the palatal shelves to simultaneously elevate (Fig. 14). The growth of the lower jaw is high relative to the remainder of the face at this stage. How the shelves elevate is unclear. However, it has been proposed that hyaluronan, a glycosaminoglycan, is a key component of this process (reviewed by Ferguson 1988). The neurotransmitter serotonin (5-HT), which is expressed in the palatal shelves prior to shelf elevation, has also been shown to mediate shelf elevation, enhancing mesenchymal contraction and migration (Fig. 14; Wee et al. 1979; Zimmerman et al. 1981, 1983). Subsequently, the palatal shelves meet and in the case of mammals, the epithelium breaks down such that mesenchymal cells from each palatal shelf intermingle. Shelf fusion starts in the centre and spreads anteriorly and posteriorly. In mice, elevation occurs at E14.5 and fusion is complete by E16.5.

Clefting of the palate may occur in at least four stages. First the shelves may not form or elevate. This may be due to a defect in development of the maxillary primordia or the mandibular primordia. If the latter does not develop properly, the tongue will not be lowered and hence will obstruct maxillary shelf elevation. This occurs in the human syndrome, Treacher Collins, and a number of mouse mutants including that in which the orphan receptor tyrosine-related protein, *Ryk*, has been genetically inactivated (Halford et al. 2000). In the *Ryk* mutant, the tongue is relatively large compared to the skeletal structures, which are in general smaller, and there is complete cleft of the palate. *Ryk* is expressed strongly in the base of the oral cavity and the tongue, and weakly in the palatal shelves between E12.5 and 13.5, being down-regulated by E14. Analysis of *Ryk* mutants at E14.5 revealed that the tongue obstructs palatal shelf elevation, suggesting that the primary defect is in a failure of lower jaw development and a relatively enlarged tongue rather than palatal shelf development and fusion. Indeed, the latter proposal is supported by the ability of the palatal shelves from *Ryk* mutant mice to fuse *in vitro* (Halford et al. 2000). Biochemical studies have shown that Ryk interacts with the ephrin receptors Ephb2 and b3 (Halford et al. 2000). Consistent with this, the double *Ephb2/Ephb3* mutants have a cleft palate phenotype similar to that of *Ryk* mutants (Table 3; Orioli et al. 1996; Adams et al. 1999).

Alternatively, the maxillary shelves may not develop properly. For example, the mouse *Pax9* mutant has cleft palate between the maxillary and palatine shelves due to failure of the palatal shelves to elevate (Fig. 14; Peters et al. 1998). Analysis has shown that at E13.5, prior to palatal shelf elevation, the maxillary shelves are abnormally shaped, suggesting that the primary defect is due to a defect in outgrowth of the primordium consistent with the expression of *Pax9* in the facial and palatal shelf mesenchyme (Peters et al. 1998). However, changes in development of the lower jaw,

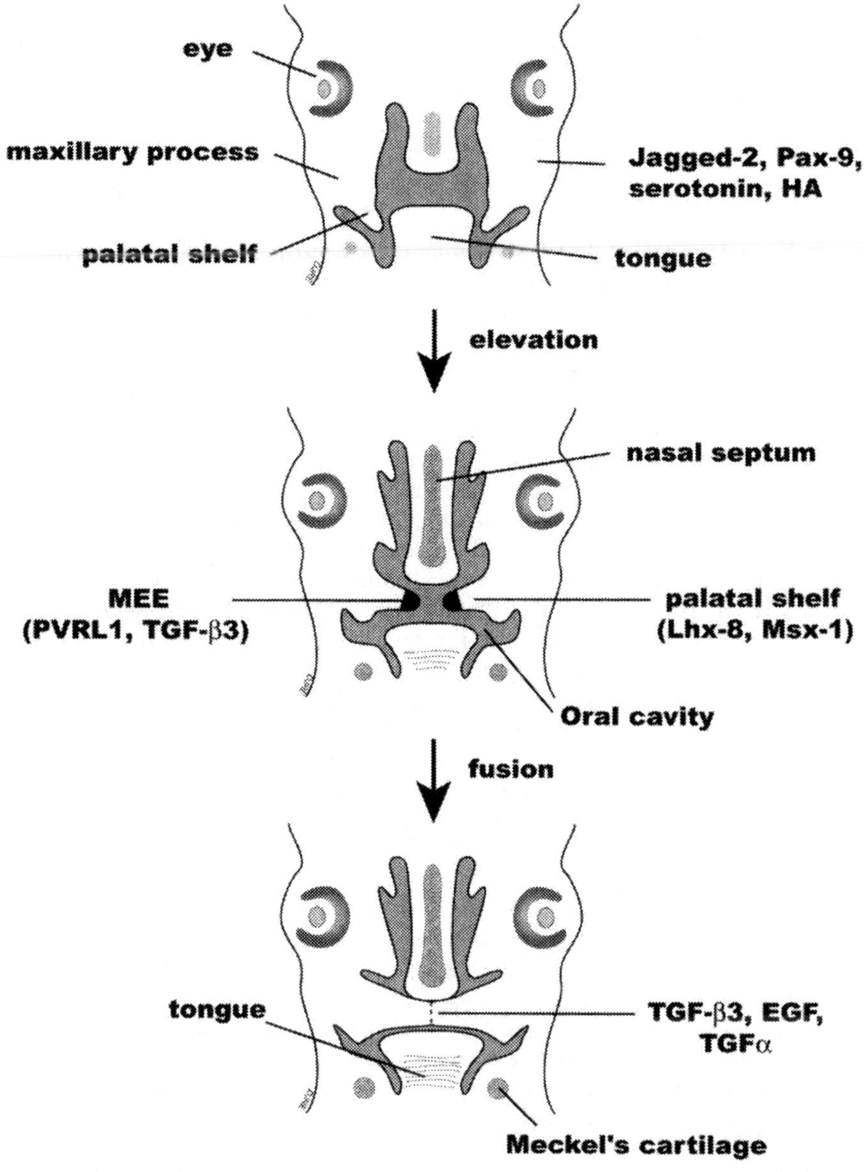

Fig. 14. Diagram of frontal sections of the head illustrating the process of secondary palatogenesis in mammals. Initially the palatal shelves project inferiorly, locating to each side of the tongue. As the tongue becomes relatively smaller and the mandibular primordia grows, lowering itself relative to the upper jaw in the process, the palatal shelves begin to elevate into a horizontal position, approaching each other and eventually fusing in the median plane. Clefting of the palate occurs when the palatal shelves fail to form, elevate, or do not meet and fuse in the midline. Some of the factors that control these processes are indicated. *Jagged-2*, *Pax-9*, *serotonin*, and *hyaluronan* (*HA*) are required for palatal shelf formation and /or elevation. *PVRL1* and TGF-β3 are expressed in the medial edge epithelium (*MEE*) whilst *Lhx-8* and *Msx-1* are expressed in the underlying mesenchyme. All of these factors are required for correct palatal shelf formation and fusion. TGF-β3, TGFα and EGF are required for the breakdown of the MEE. See Sect. 5.3 and Tables 3 and 5 for more information

which is also misshapen, may contribute to the final cleft palate phenotype (Peters et al. 1998). A similar cleft palate phenotype is seen in the *jagged 2* or *Sm* mutant. Again the palatal shelves do no elevate but in this case, the shelves fuse with the tongue (Fig. 14; Sidow et al. 1997; Jiang et al. 1998a; Luo et al. 1997).

Following palatal shelf elevation the shelves must meet and fuse. Defects in this process can occur because the palatal shelves are physically too small to meet as occurs in the *Msx1* mouse mutant, which has a defect in epithelial-mesenchymal interactions and hence outgrowth (Fig. 14; Satokata and Maas 1994). A recent paper has investigated the role of Bmp4-Msx1 interactions during palatal shelf development. The study shows that *Msx1* controls *Bmp2/4* expression in the mesenchyme, together with *Shh* expression in the MEE. Furthermore, Bmp4 can rescue the palatal shelf defects in *Msx1* mutant mice (Zhang et al. 2002. Development 129: 4135–4164). A cleft palate phenotype is also seen in humans following mutation in *MSX1* (van den Boogaard et al. 2000). Alternatively, the shelves meet but they fail to fuse due to maintenance of the medial edge epithelium (MEE). The latter is composed of two layers, an outer flattened cell layer, the peridermal layer, and an inner basal cell layer, which are separated from the underlying mesenchyme by a basal membrane. The MEE cells cease to proliferate approximately 24–36 h prior to adhering (Hudson and Shapiro 1973; Pratt and Hassell 1975; Pratt and Martin 1975). Following contact, the basal membrane is degraded whilst the peridermal layer is sloughed off and/or undergoes resorption. The remaining epithelial cells make contact via desmosome and filopodia-like structures, ultimately being removed by a combination of cell death, migration, and epithelial-mesenchymal transformation (Pratt and Martin 1975; Ferguson 1988; Fitchett and Hay 1989; Griffith and Hay 1992; Martínez-Álvarez et al. 2000a,b; Mogass et al. 2000; Cuervo et al. 2002). The increased adhesive ability of the palatal shelves is associated with cell surface changes in glycoproteins, such as an increase in ConA binding and glycoprotein-rich material, which occurs at the onset of fusion and is crucial (Pratt and Hassell 1975). Thus, inhibition of glycosaminoglycan and glycoprotein synthesis inhibits palatal shelf fusion (Pratt and Hassell 1975; Greene and Pratt 1977).

In mice, cell death in the MEE occurs in the posterior palate before shelf contact, whereas cell death in the anterior palate is contact-dependent (Cuervo et al. 2002). Cell death is mediated by retinoid signalling, and its importance in generating a functional palate is illustrated by the knockout of *Apaf1*, a factor that controls apoptosis (also see Sect. 2.1). In these mutants, the levels of apoptosis are decreased and there is midline facial cleft, which includes cleft palate where the palatal shelves meet but fail to adhere (Cecconi et al. 1998). In the absence of cell contact or fusion, as occurs naturally in the chick, the epithelium becomes stratified and keratinised.

TGF-β3, which is expressed in the MEE in mice and rats, appears to be a key molecule in fusion of the palatal shelves (Fig. 14; Fitzpatrick et al. 1990; Pelton et al. 1990). Loss of TGF-β3 function using neutralising antibodies, antisense oligos, or by gene inactivation in mice results in cleft palate (Brunet et al. 1995; Kaartinen et al. 1995, 1997; Proetzel et al. 1995). In *TGF-β3* null mice, palatal shelf contact occurs but there is little apoptosis, and the basement membrane together with the MEE is retained (Kaartinen et al. 1997; Martínez-Álvarez et al. 2000b). Epithelial migration also appears to be defective as suggested by the absence/reduction in the epithelial triangles at the junction of the midline seam and nasal or oral epithelium (Taya et al. 1999). Other analyses have shown that TGF-β3 is necessary for formation of the glycoprotein-coated filopodia, and promotes fusion in a dose-dependent manner (Taya et al. 1999). In chicks the palatal shelves do not touch and will only undergo partial fusion when placed in direct contact *in vitro*. Normally in chicks there is continued proliferation and no cell death of the MEE (Greene et al. 1983). This failure to undergo fusion even when placed in contact appears to be due to the lack of TGF-β3 and indeed, fusion can occur when TGF-β3 is added (Sun et al. 1998; Taya et al. 1999). The target of TGF-β3 appears to be the MEE cells themselves as inferred from the expression of the TGFβ receptors, R1 and RII, in the midline seam (Taya et al. 1999). The related factor TGF β1 is also expressed in the MEE whilst TGF-β2 is expressed in the mesenchymal cells (Pelton et al. 1990; Taya et al. 1999). These factors can also promote palatal shelf fusion but to a far lesser extent (Taya et al. 1999).

Another growth factor expressed in the MEE is TGFα, which is present together with its receptor EGFR (Fig. 14; Miettinen et al. 1999). Loss of function of *EGFR* results in cleft palate where the palatal shelves have not totally fused and epithelial remnants remain. This may be related to decreased metalloproteinase activity. Delayed growth of Meckel's cartilage and thus of the lower jaw may also be contributory factors (Miettinen et al. 1999).

Finally, muscle development also appears to be important for correct palatal development. The craniofacial muscles, including the tongue muscles, are functional prior to palatal shelf closure, suggesting that their activity may contribute to successful palatal shelf closure (Wragg et al. 1972). Indeed, following gene inactivation of *Hoxa2*, the hyoid muscles, with the exception of the thyrohyoideus, and the tongue muscles, with the exception of genioglossus, do not "migrate" properly and/or are abnormally inserted (Barrow and Capecchi 1999). The palatal defect that occurs in some of the *Hoxa2* mutants is apparently not due to the skeletal changes and has been precisely correlated with the failure of the hypoglossus muscle to attach to the greater horn of the hyoid (Barrow and Capecchi 1999).

In addition to genetic factors, environmental insults can result in cleft palate. For example, levels of vitamin A derivatives and GABA signalling are critical, with both deficiencies and excess resulting in cleft palate. Loss of GABA signalling in mice and excess in humans, by administration of drugs such as the antidepressant diazepam, increases the risk of cleft palate (Miller and Becker 1975; Wee and Zimmerman 1983; Asada et al. 1997; Condie et al. 1997; Homanics et al. 1997). Similarly excess and loss of retinoid signalling results in cleft palate. The teratogenic effects of retinoids are mediated, at least in part, by the retinoid receptor, RXRα (Sucov et al. 1995; Nugent et al. 1999). Excess retinoids can affect palatal development in several ways. First, retinoids can affect neural crest survival and/or migration (Moro-Balbas et al. 1993; Lee et al. 1995; Salvarezza and Rovasio 1997). Second, retinoid signalling has been shown to maintain proliferation and survival of the MEE which then becomes stratified, although more recent studies have shown that excess retinoids can also induce cell death (Abbott and Pratt 1987a,b, 1991; Watanabe et al. 1988; Abbott 1995; Cuervo et al. 2002). Finally, the environmental toxin, 2,3,7,8-tetrachlororidibenzo-p-dioxin (TCDD) inhibits MEE degeneration (Abbott 1995).

6 Skeletal Development

There are two types of bone in the head: endochondral bone, which is formed from a cartilaginous precursor, and membrane (or intramembranous) bone, which arises by direct differentiation of mesenchymal cells into osteoblasts. Endochondral bone forms the chondrocranium – the base of the skull, Meckel's and hyoid cartilages in the first and second arches respectively, the nasal septum and capsule and the otic capsule. The other bones in the head are formed by intramembranous ossification, which start as ossification centres which spread to form the bone. Secondary cartilages, which are again unique to the head in higher vertebrates, arise in the periosteum of membrane bones in response to movement (reviewed by Fang and Hall 1997). The most important of these is the condylar cartilage on the mandible bone, which persists post-natally. Secondary cartilage has distinct histological properties and functions from primary cartilage. The former is formed in response to pressure changes and is essential for normal growth of the face, whereas the latter is resistant to pressure and initially provides a structural scaffold, and later a growth site in the case of the nasal septum and cranial base, for the developing head.

6.1 Induction of Skeletal Structures

Many of the bones of the head arise by inductive interactions. The hindbrain and notochord induce parachordal cartilage, the basioccipital and basisphenoid (Holtfreter 1968), whilst the posterior hindbrain and prosencephalon induce the squamosal, occipital, supra-occipital, parietal, and frontal bones (Showing 1968). Again emphasising the link between brain and skull development in exencephalic *Cart1* mouse mutants, but not in the *Cart1* mutants where the neural tube has closed, the cranial vault is absent (Zhao et al. 1996). Some membrane bones are initiated at the sites of bifurcation of nerves, such as the ossification centre of the mandibular bone, which arises at the bifurcation of the mandibular nerve into the inferior alveolar and lingual nerves. The relationship between nerve bifurcation and ossification is unclear but it is possible that the same factors control both processes. In addition, the epithelium has been implicated for the acquisition of chondrogenic potential in neural crest cells and later, for patterning parts of the chondrocranium (Epperlein and Lehmann 1975; Bee and Thorogood 1980; Graveson and Armstrong 1987; reviewed by Hanken and Thorogood 1993). For example, the otic placode induces the otic capsule (Frenz and van de Water 1991). Epithelial signals are also required for the induction of membrane bone differentiation in the maxillary and mandibular arches (Bee and Thorogood 1980; reviewed by Hall and Miyake 1995, 2000).

Once the membrane bones have been induced, they grow and meet at sutures, mesenchymal gaps between the developing bones, which are important for growth. Fontanelles are the unossified regions between more than two skull vault bones. In the human skull, sutures typically form between the frontal bones (metopic, or interfrontal), the parietal bones (sagittal suture), the frontal and parietal bones (paired coronal sutures), the supraoccipital and parietal bones (lambdoid sutures), and the parietal, temporal, and sphenoid bones (squamosal sutures). The signals and tissues that induce suture formation are unknown (reviewed by Opperman 2000); however, the dura mater is needed to initially maintain the suture once it has formed (Opperman et al. 1993). In the absence of the dura mater, newly formed sutures are unable to remain open (Opperman et al. 1995, 1998). Other sutures such as that between the bones in the maxilla are not in contact with the dura mater, and the signals/tissues that maintain them have not been identified. However, ablation experiments have shown that the ectoperiosteal layer is not required for the suture to remain open (Opperman et al. 1994).

The epithelial and neuroectodermal signals that have been proposed to induce bone and cartilage formation include Fgf2 and TGFβ. Fgf2 can induce chondrogenesis and membrane bone formation in migratory cranial neural crest cells, whilst TGFβ together with Fgf signalling has been implicated in the induction of the otic capsule in post-migratory neural crest cells (Frenz et al. 1992, 1994; Sarkar et al. 2001). In addition, Shh must be a candidate as it is expressed in sites of inductive interactions, including the facial endoderm, and it specifies the sclerotome and hence skeletal development in axial regions (Chiang et al. 1996). As in other regions of the body, Sox-9 appears to be the key down-stream mediator in determining chondrogenic commitment. A recent study has shown that the activity/effect of Sox-9, which is expressed in the migrating cranial neural crest, is initially counteracted by the co-expression of Msx2, at least in a sub-population of cells (Takahashi et al. 2001). Other transcription factors that are important include Dlx5 and Gsc, although they may have a role in morphogenesis and skeletal differentiation rather than the initial induction and commitment. *Dlx5* is expressed in the nasal placode and pit, regulating *Gsc* expression in the underlying mesenchyme. Loss of function of either results in hypoplasticity of the nasal capsules, which are induced to form by the nasal epithelium (Rivera-Perez et al. 1995; Yamada et al. 1995, 1997; Acampora et al. 1999; Depew et al. 1999).

6.2 Skeletal Growth and Differentiation

A number of growth and transcription factors that control skeletal development have been identified. Overall the analysis has shown that the development of the skeletal structures in the different regions of the head is controlled by a distinct set of signalling interactions and thus, a mosaic of signalling networks combines to co-ordinate the intricate development of the skull.

Alx4 is a homeobox-containing gene related to the *Drosophila* gene *aristaless*, which is expressed in mesenchymal condensations including the nasal and maxilla ossification centres, and later the undifferentiated layers of the bone and sutural mesenchyme (Hudson et al. 1998; Mavrogiannis et al. 2001). Loss of function of *Alx4* in mice affects the development of only one bone, the parietal, which is smaller (Qu et al. 1997). This defect appears to be a developmental delay. Three distinct mutations, two nonsense and one missense in the homeodomain, have been identified in humans which result in a deficiency of mineralisation and symmetric parietal formina (PFM) (Wu et al. 2000; Wuyts et al. 2000; Mavrogiannis et al. 2001). The latter is the result of ossification defects in the parietal bones. The openings that are formed are usually symmetrical to the sagittal suture and are variable in size but they do tend to decrease with time. These lesions in the skull are not usually clinically significant, although if they are very large they may need surgical intervention. All the *ALX4* mutations are predicted to be a loss of function. Loss of function of the related gene, *Cart1*, which is also expressed in ossification centres, also affects skeletal differentiation in the mouse (Zhao et al. 1996; Hudson et al. 1998). The double *Cart1/Alx4* mouse mutant has a more enhanced phenotype, suggesting a redundancy of function in the areas of overlapping gene expression (Qu et al. 1999).

Haploinsufficiency of *MSX2* in humans has also been implicated in PFM (Wilkie et al. 2000). A similar phenotype is seen following gene inactivation of *Msx2* in mice, which have calvarial abnormalities including smaller and misshaped bones resulting in an enlarged foramen (Satokata et al. 2000). This is due to decreased proliferation of the undifferentiated osteogenic precursors. There is also delayed ossification in *Msx1* mutant mice resulting in an enlarged anterior fontanelle (Satokata and Maas 1994). Like *Msx2*, *Msx1* is expressed in skeletal precursors. The double *Msx1/-2* mouse mutant is characterised by the absence of membrane bones at E18.5, whilst the endochondral bones remain relatively unaffected (Satokata et al. 2000). This shows that there is a redundancy between Msxl and -2 function during intramembranous development and emphasizes that the processes that control endochondral and membrane bone formation are distinct.

Another homeobox-containing gene required for skeletal development is *Bapx1*, which is required for development of some of the bones derived from the mesoderm. Following loss-of-*Bapx1* function in mice, the basisphenoid and exoccipital are malformed whilst the supraoccipital is absent (Lettice et al. 1999; Tribioli and Lufkin 1999; Akazawa et al. 2000). Analysis has shown that the early steps of differentiation are affected with the mesodermal cells failing to undergo the prechondroblast to chondrocyte transition, and instead undergoing apoptosis (Tribioli and Lufkin 1999; Akazawa et al. 2000).

In contrast to the *ALX4* and *MSX2* mutations, which are characterised by a deficiency in mineralisation, craniometaphyseal dysplasia, an autosomal syndrome, is characterised by thickening of the craniofacial bones, together with abnormal growth of the long bones with epiphyseal flaring (Nürnberg et al. 2001). The expansion of the cranial bones can trap the facial, vestibulocochlear, and optic nerves, ultimately resulting in hearing loss and facial palsy or paralysis. This syndrome is due to a nonsense mutation in *ANKH*, a transmembrane protein involved in the transport of inorganic pyrophosphate, which normally inhibits calcification, bone mineralisation, and remodelling (Nürnberg et al. 2001). These phenotypes are not seen in cri-du-chat syndrome, in which the *ANKH* locus has been deleted, and hence are predicted to be due to gain of function. A mouse *Ankh* mutant also exists. This mouse mutant, which is predicted to be due to loss of function, does not have craniofacial defects but develops arthritis within six months of birth due to the formation of hydroxapaptite crystals in the joint region (Ho et al. 2000). *DLX3* mutations in humans cause the autosomal dominant syndrome, trichodento-osseous syndrome, where the cranial bones are also thicker (Price et al. 1998). This can result in narrowing or obliteration of the frontal and mastoid sinuses. The teeth are also affected. The malformation of the bones presumably reflects a failure in bone remodelling.

Finally, the most prevalent skeletal disorder is craniosynostosis, premature fusion of the sutures, which is a component of over 100 syndromes in humans (reviewed by Morriss-Kay et al. 2001; Wilkie et al. 2001). They are all characterised by the premature fusion of one or more of the cranial sutures, which lead to skull deformation and dysmorphic facial features of variable severity. The abnormalities of the skull usually are manifested as a tall forehead, with widely spaced and prominent eyes (proptosis) and mid-face hypoplasia. The gene expression data and genetic studies in both mice and humans has shown that the signalling processes that control suture development are not the same in each suture. These variations are reflected in the different incidences of synostosis of single sutures – the sagittal suture is the most commonly affected (50%–60%), followed by one or both coronal sutures (17%–30%). Most of these craniosynostosis syndromes are inherited as autosomal dominant traits, and genetic studies have now identified that many of these syndromes are caused by mutations in *fibroblast growth factor receptors 1, 2, 3* (*FGFR1–3*), *MSX2,* or *TWIST* (reviewed by Morriss-Kay et al. 2001; Wilkie et al. 2001). The mutations in *FGFR* and *MSX2* have been shown to be gain-of-function, whereas the mechanism of the defect resulting from the *TWIST* mutations is unidentified (Table 5).

7 Muscle Development

Compared to the development of the trunk and limb muscles, where the complex set of tissue and molecular interactions has been unravelled in great detail, there is an extreme paucity of data about craniofacial myogenic differentiation. For example, in the limb we know that ectodermal signals, mediated by Bmps, maintain proliferation and prevent myogenic differentiation whilst the signalling factors scatter factor and Fgfs control migration and differentiation (reviewed by Buckingham 2001). However, this knowledge cannot be used to dissect the cascade of craniofacial myogenesis as these cells usually do not express gene markers, such as *Pax3* and *Lbx1*, which mark uncommitted myogenic cells in other regions of the body, suggesting that their development is distinct. Expression of the myogenic determination factor, *Myf5*, is also under the control of distinct promoters in the head and trunk (Hadchouel et al. 2000; Summerbell et al. 2000; Carvajal et al. 2001). The proposed differences have been further supported by studies in which undifferentiated craniofacial myogenic cells have been grafted into the trunk and *vice versa*, where the cells have been shown to maintain their original identity (Hacker and Guthrie 1998; Mootoosamy and Dietrich 2002). However, whether these studies provide conclusive proof that cranial and trunk myogenic cells are intrinsically different is still far from clear. These grafting experiments transpose relatively large pieces of tissue and hence the premyogenic cells are grafted with surrounding tissue. As with the cranial neural crest there may be a community effect or the surrounding mesoderm may maintain the appropriate genetic code (see Sect. 5.1). Hints at this come from experiments by Hacker and Guthrie (1998) – where the somitic cells become detached from the main graft of cells, they switch off the expression of *Pax3* in accordance to the local environment. These questions will only be answered by clonal analysis in which individual myogenic cells are transferred into an ectopic environment.

7.1 Origins and Migration

Unlike other regions of the body where the musculature arises from the somites, the origins of the craniofacial muscles is remarkably heterogeneous, receiving contributions from the prechordal mesoderm and unsegmented cranial paraxial mesoderm, in addition to the segmented (somitic) mesoderm, which gives rise to the tongue, laryngeal, and pharyngeal musculature (Fig. 15; Hazelton 1970; Wachtler and Jacob 1986; Noden 1983a; Trainor and Tam 1995; Mackenzie et al. 1998; Huang et al. 1999, 2000; also see Sect. 3.2). A map of the origins of head muscle precursors and their destinations is shown in Fig. 15. The extraocular muscles have myogenic precursors

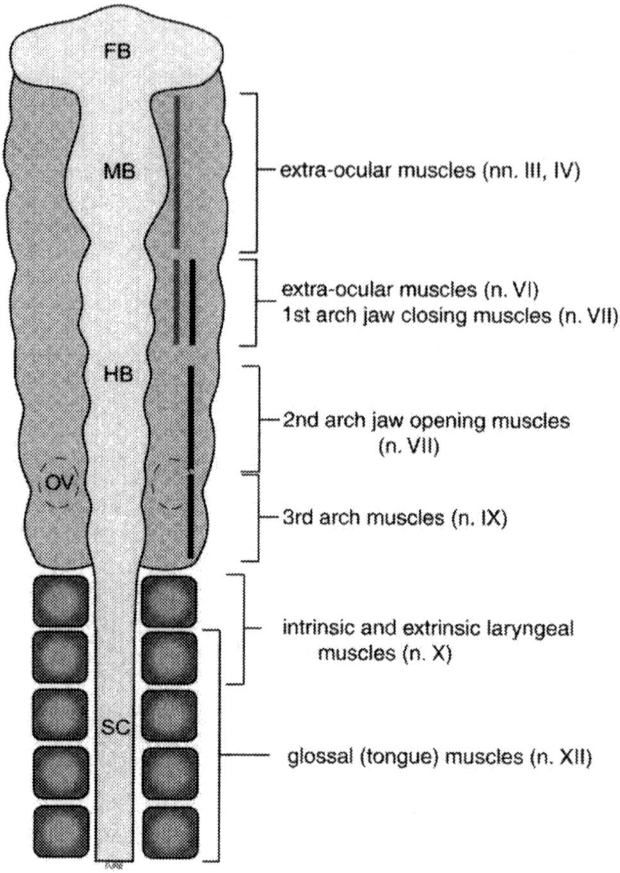

Fig. 15. Illustration of the origins of the precursors of craniofacial muscles within the cranial paraxial mesoderm. Precursors of the extraocular muscles (*grey bars*) are located more medially within the mesoderm, and the precursors of the branchiomandibular muscles (*black bars*) originate more laterally. The laryngeal and glossal muscles arise from somitic mesoderm. *FB*, forebrain; *MB*, midbrain; *HB*, hindbrain; *SC*, spinal cord; *OV*, otic vesicle. (Based on the results of Noden 1983a; Couly et al. 1992; Noden and Evans, unpublished data)

that arise from both the prechordal mesoderm and the unsegmented paraxial mesoderm (Noden 1991a). The former are innervated by cranial nerve III (occulomotor) while the latter are innervated by cranial nerves IV and VI (trochlear and abducens). The other craniofacial muscles, including those of the jaws, larynx, and tongue, also arise in register with the origin of the motor axons (Huang et al. 1999). The origin of the connective tissue for the head muscles is also heterogeneous with both neural crest and mesodermal contributions (Noden 1983a).

The migration of the myogenic precursors of the head is different from the migration of the limb myogenic precursors. In the limb the myogenic cells migrate individually, while the myogenic cells of the head move en masse into the appropriate future branchial arch at the same time as the neural crest, and as the surface ectoderm is expanding (Noden 1984). This migration is a very early event in the development of the embryo. In the chick embryo, movement of the myogenic cells and neural crest begins at stage 10 and is mainly over by stage 12–13 (Noden 1988). In arches 1 and 2, the neural crest cells tend to migrate superficially to the myogenic cells separating them from the surface ectoderm (see Fig. 7). As a result the neural crest cells initially encapsulate a mesodermal core containing the myogenic precursors, although later the neural crest cells infiltrate into the mass of myogenic precursors where they differentiate into the craniofacial muscle connective tissues (McClearn and Noden 1988). Throughout these morphogenetic events, the branchial arch myogenic cells also remain in close contact with the motor axons that will innervate them (Noden 1991a).

In contrast to the other muscles, the tongue muscles and the muscles of the lateral rectus complex (those innervated by abducens) actively move relative to other tissues. The tongue musculature actively migrates and as such expresses molecular markers typical of migrating myogenic cells such as *Pax3* and *Lbx1* (MacKenzie et al. 1998; Brohmann et al. 2000). In contrast, the lateral rectus muscle moves from its origin adjacent to the rostral hindbrain towards the developing eye, but how this is achieved is unclear. It may be an active migration or a result of cranial flexing (Wahl et al. 1994). The precursors of the lateral rectus, and the chick-specific muscles pyramidalis and quadratus, are also unique amongst the craniofacial muscles as they begin to differentiate before they reach their final destination (Wahl et al. 1994).

The tongue myogenic cells arise from the ventral edges of the occipital somites and condense together to form the hypoglossal cord, which then migrates into the hypopharyngeal region, first along the base of the sixth pharyngeal arch to the ventral tip. Then the myogenic cells sweep anteriorly underneath the endoderm of the more rostral arches (Mackenzie et al. 1998; Brohmann et al. 2000; Uchiyama et al. 2000). This pathway is the same taken earlier during development by the hypoglossal nerve, which extends into the tongue anlagen. The neural crest cells arising from this region initially also migrate along this route prior to myogenic migration, suggesting that the neural crest cells and/or the hypoglossal nerve may provide the ques for myogenic migration. However, as neural tube ablation, which removes both the neural crest and hypoglossal nerve, does not affect the development of the tongue muscles, this suggests that environmental signals, presumably similar to those that control neural crest and axon guidance (see Sect. 4.3) must be important (MacKenzie et al. 1998; Huang et al. 1999). Unlike in the limb, scatter factor does not appear to play a role – judged by its expression pattern where it is only found in the distant first arch (MacKenzie et al. 1998). Factors that control later stages of myogenic cell migration within the developing tongue itself include gelatinase A and stromelysin-3 (Chin and Werb 1997).

7.2
Muscle Differentiation

As with all muscles, commitment to myogenesis is marked by the expression of members of the myogenic determination factor family, which include the basic helix-loop-helix transcription factors *Myf5* and *MyoD* (Fig. 16; Hacker and Guthrie 1998; Noden et al. 1999). A recent study has shown that inactivation of both *capsulin* and *MyoR*, factors related to the myogenic determination factors, affects the development of some of the masticatory muscles together with the diaphragm musculature, show-

progenitor Cells

Myf-5 expression induced

MyoD expression induced and Myf-5 expression maintained

condensation of myogenic cells

fusion of myogenic cells and expression of MyHCs

Fig. 16. Schematic diagram of the formation of craniofacial muscle fibres and the expression of muscle-specific genes. *Myf5* appears to be the first muscle regulatory factor to be expressed by head muscle precursors, followed soon after by *MyoD* expression. After condensing, some muscle precursors begin to fuse into primary myotubes, which subsequently express myosin heavy chain isoforms

ing that unique regulation and development of craniofacial muscles compared to the rest of the body (Lu et al. 2002, Science 298: 2378–2381). The majority of the head muscles show an exceptionally prolonged period of *Myf5* and then *MyoD* expression compared to other regions of the body. This is typically over 40 h, with the muscles in the tongue expressing these factors for 4 days before terminally differentiating (Noden et al. 1999). It has been proposed that signals from the neural crest cells or the pharyngeal endoderm might delay the terminal differentiation of the craniofacial myogenic cells. However, as in the limb, ectodermal signals may also play an inhibitory role. In the tongue, the onset of differentiation is associated with TGFα and EGF, which promote myogenic differentiation (Yamane et al. 1997, 1998). Another factor implicated in myogenic differentiation is scatter factor. Its expression colocalises with *MyoD*, which is switched on after *scatter factor* (MacKenzie et al. 1998). Furthermore, following loss of function of *scatter factor* or of its receptor, *c-met*, there is a reduction in the number of tongue myogenic cells (Bladt et al. 1995). This is not due to a migration defect but to abnormalities in later steps of differentiation. If scatter factor is required for myogenic differentiation, this would be very surprising – in the developing limb bud, scatter factor has the opposite effect and prevents differentiation (reviewed by Buckingham 2001). This apparent discrepancy may reflect the uniqueness of the craniofacial musculature. Other hints that the craniofacial musculature may have a different story to tell come from the expression of the Wnt antagonist, *Sfrp2* (Ladher et al. 2000b). In the limb, *Sfrp2* is expressed in undifferentiated myogenic precursors, and like other Wnt antagonists in the somites, appears to prevent terminal differentiation (Ladher et al. 2000b). Yet in the face, *Sfrp2* expression is associated with differentiating muscles (Borello et al. 1999; Ladher et al. 2000b).

The initial commitment to myogenesis is followed by the activation of muscle-specific differentiation genes, which will produce the muscle proteins. These include the expression of slow and fast myosin heavy chains, which determine how a muscle will function (Table 4). Fast fibres have glycolytic activity, contract rapidly and are needed for movement, whilst slow fibres have oxidative metabolism and contract slowly, being

Table 4. Summary of the mammalian myosin heavy chain isoforms found in developing and adult craniofacial muscles

Mammalian myosin heavy chain isoforms	
Embryonic isoforms	
MyHC-Emb	Isoform expressed in all developing fibres
MyHC-Peri	Isoform expressed in future fast fibres
Adult isoforms	
MyHC-β	Isoform expressed in developing and adult slow fibres
MyHC-α	Isoform expressed in some head muscles
MyHC-IIa	Adult fast isoform
MyHC-IIx/d	Adult fast isoform
MyHC-IIb	Adult fast isoform
MyHC-Eo	Isoform expressed in extraocular and some other head muscles
MyHC-IIm	Fast isoform expressed in some jaw muscles (carnivores)

mainly required for posture. Typically, muscles are characterised by a fibre-type phenotypic range of fast type IIa and IIb and the slow type I, but some craniofacial muscles express unique myosin isoforms (Table 4). The masticatory muscles of carnivores express a tissue-specific myosin which resembles the embryonic myosin seen in all muscles as they differentiate but which is usually down-regulated as the muscles start to express the adult isoform (Shelton et al. 1988; Sciote et al. 1994, 1995; Soussi-Yanicostas et al. 1990). The extraocular muscles also have a distinct profile of myosin expression (Porter et al. 2001). They express an extraocular-specific myosin isoform, and unlike the majority of muscles, around 20% of their fibres remain multiply innervated (Hoh 1992; Korfage et al. 2001). In mice, the homeobox-containing gene *Engrailed-2* (*En-2*), which is expressed in the first arch myogenic precursors, later becoming confined to a restricted subset of muscles (masseter, temporalis, medial, and lateral pterygoids), has been shown to control fibretype specification of the masseter muscle in a sexually dimorphic manner (Degenhardt et al. 2002). Overexpression of En-2 increases the percentage of intermediate fibres, i.e. those possessing both slow and fast characteristics, and decreases the cross-sectional size of the glycolytic fibres in both males and females. However, loss of function of *En-2* decreases the percentage of intermediate fibres in females only (Degenhardt and Sassoon 2001).

There are several waves of myogenesis: the first wave of differentiation forms the primary fibres that usually lay down the pattern of fast and slow fibres of the secondary fibres during fetal development and the myoblasts that arise from the satellite cells in the adult (Robson 1993). The exceptions to the rule include the primary fibres within the oculorotatory muscles, which are slow. In contrast, the secondary fibres consist of both slow and fast fibres with the slow fibres forming a rim around the perimeter of the muscles, closest to the bony orbit (Marcucio and Noden 1999). Another exception is the large mandibular depressor muscle of the second arch, which has no slow primary myofibres but which becomes mixed during secondary fibre myogenesis (Marcucio and Noden 1999; reviewed by Wigmore and Evans 2002).

The formation of the individual muscles is accomplished by the division of the muscle plates (Fig. 17; McClearn and Noden 1988; Robson 1993). By stage 32 in the

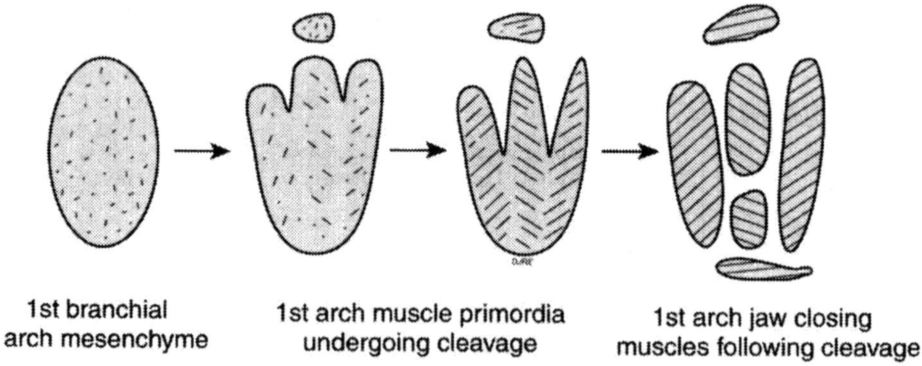

Fig. 17. Diagram of the cleavage of branchial arch mesenchyme first into the primordia of and subsequently the individual muscles of the first branchial arch. (Based on the studies of McClearn and Noden 1988 and unpublished diagrams of Drew Noden)

chick head, most of the muscles have fully segregated from the muscle plates and the fast and slow fibre pattern has been established. As the myogenic cells begin to differentiate, within the muscle plate, the muscle fibre orientation for each muscle is established and even presages the division of the muscle plates into the individual muscles (Fig. 17; McClearn and Noden 1988; Robson 1993).

8 Human Craniofacial Syndromes

There are many syndromes which include craniofacial abnormalities. The defects may be restricted to specific regions of the head, for example the skull as occurs in craniosynostosis or specific regions of the face such as the first branchial arch as occurs in Treacher-Collins syndrome. In addition, as the heart and circulatory system also require cranial neural crest cells, the craniofacial abnormalities may be linked to cardiac and aortic arch defects reflecting the common origin of these structures as seen in DiGeorge syndrome. As neural and head development are interdependent, these syndromes may also be linked to neural defects. The most striking examples of these are exencephaly and holoprosencephaly, which were discussed in Sects. 2.1 and 2.2. In Table 5 we have summarised the human syndromes where the gene mutations responsible have been identified. In the following sections we describe three syndromes – Riegers, Treacher-Collins, and DiGeorge – where there has been a large advance of knowledge over the past few years in both the molecular and tissue bases of the syndrome.

8.1 Riegers Syndrome

Mutation of the bicoid-type homeobox-containing protein Pitx2 (also known as Reig, Ptx2, ARP1, and Otxl-2) results in the autosomal dominant Riegers (or Axenfeld-Rieger) syndrome type 1 predominantly characterised by anterior eye, dental, and umbilical stump defects (Jorgenson et al. 1978; Semina et al. 1996; Flomen et al. 1998). The ocular abnormalities include the cornea, iris stroma, and trabecular network, and result in glaucoma in over 50% of patients. There may be ectopic pupils or the pupils may be displaced. The dental defects range from abnormally shaped teeth and missing upper incisors to a total lack of teeth. The umbilical stump protrudes because the periumbilical skin does not involute. In the worst cases, this can be associated with omphalocele, where the abdomen does not close. Less frequently, there are also cardiac, pituitary, and limb abnormalities (Sadeghi-Nedjad and Senior 1974; Mammi et al. 1998). However, mutations in *PITX2* only account for 40% of Riegers syndrome patients, and other loci are involved. These may encode homologues of or downstream targets of Pitx2 (reviewed by Amendt et al. 2000).

Pitx2 shows a highly specific pattern of expression during development and is found in all the affected organs. Thus, *Pitx2* transcripts are found in the periocular mesenchyme, the dental lamina, heart, limb, and Rathke's pouch, which gives rise to the anterior and intermediate pituitary (Muccielli et al. 1996; Semina et al. 1996; Gage

Table 5. The human syndromes which have craniofacial defects where the gene mutation(s) responsible has been identified[a]

Syndrome	Gene	Phenotype: Craniofacial	Phenotype: Other	Mechanism of mutation	Omim reference	References
Midline defects						
Holoprosencephaly	SHH (growth factor)	Hypo- or hypertelorism Cleft lip and palate Microcephaly Single central incisor Flat or absent nose Cyclopia Philtrum hypoplasia	Mental retardation Corpus callosum agenesis Developmental retardation Pituitary gland hypoplasia Inferior iris colobomata Anosmia	Loss of function	236100 142945 600725	Belloni et al. 1996 Nanni et al. 1999, 2001 Odent et al. 1999 Roessler et al. 1996, 1997
	SIX3 (transcription factor)	Hypotelorism Midline cleft and cleft lip Nasal bones absent Microcephaly	Mental retardation Micro-opthalmic Iris coloboma	Mutations in homeodomain Loss-of-function?	157170 603714	Wallis et al. 1999
	TGIF (transcriptional repressor)	Hypo- or hypertelorism Midline cleft and cleft lip Midface hypoplasia Microcephaly	Developmental retardation Mental retardation Corpus callosum agenesis	Loss-of-function Normally inhibits retinoid and nodal signalling?	142946 602630	Gripp et al. 2000
	ZIC2 (transcription factor)	Minor face defects Microcephaly	Developmental retardation Corpus callosum agenesis	Loss-of-function or dominant-negative	603073	Brown et al. 1998, 2001
Opitz	MID1 (intracellular protein)	Cleft palate and lip hypertelorism	Hypospadias Mental retardation Imperforate anus Laryngeal cleft Tetralogy of Fallot	Function unknown Prevents association with microtubules	300000	Cainarca et al. 1999 Dal Zotto et al. 1998 Liu et al. 2001 Quaderi et al. 1997 Schweiger et al. 1999
Exencephaly						
Knobloch syndrome	α1(XVIII) collagen short chain	Flat nasal bridge Midface hypoplasia Hypertelorism	Exencephaly – occipital Spina bifida Kidney defects Eye defects	Truncated protein	267750 120328	Sertié et al. 2000

Table 5. *Continued*[a]

Syndrome	Gene	Phenotype		Mechanism of mutation	Omim reference	References
		Craniofacial	Other			
Facial defects						
Alagille	JAGGED-1 (growth factor)	Hypertelorism Deep-set eyes Nose defects – bulbous tip Flat face Short philtrum Pointed chin Prominent forehead	Liver Heart Eye Vertebrae	Haploinsufficiency	118450 601920	Crosnier et al. 2001 Oda et al. 1997 Pilia et al. 1999 Yuan et al. 1998
Rieger Syndrome, type I	RIEG (PITX2) Transcription factor	Maxillary hypoplasia Loss of or misshapen teeth	Eye Hypospadias Periumbical plate does not involute	Haploinsufficiency or dominant-negative	180500 601542	Kulak et al. 1998 Semina et al. 1996 (see Sect. 8.1)
Char syndrome	TFAP2B (transcription factor) (Ap2ß)	Face defects	Hand defects Heart defects	Dominant-negative	169100 601601	Satoda et al. 2000 Zhao et al. 2001
Ectrodactyly ectodermal dysplasia/facial cleft (EEC3)	p63	Tooth defects Cleft lip and palate	Limb Skin, hair, nails Urogenital Lungs	Gain-of-function or dominant-negative	604292	Ianakiev et al. 2000 van Bokhoven et al. 2001
Cleft lip/palate/ Ectodermal dysplasia1	PVRL1 (nectin-1)	Cleft lip and palate	Hand defects Skin defects Mental retardation	Dominant-negative or Gain-of-function	225060 600644	Sozen et al. 2001 Suzuki et al. 2000
Congenital hypothyroidism	FKHL15 Titf-2 (transcription factor)	Cleft palate	Thyroid Choanal atresia Spiky hair	Loss of DNA binding and transcriptional activity	218700 241850	Clifton-Bligh et al. 1998

Table 5. *Continued*[a]

Syndrome	Gene	Phenotype: Craniofacial	Phenotype: Other	Mechanism of mutation	Omim reference	References
Greig cephalopoly-syndactyly syndrome	GLI3	Hypertelorism High prominent Forehead Macrocephaly Broad nasal root	Limb defects	Haploinsufficiency	175700 165240	Kalff-Suske et al. 1999 Vortkamp et al. 1991 Wild et al. 1997
Oligodontia	PAX9	Permanent molars absent Other teeth may be affected		Unknown Mutations in paired-domain	604625 167416	Stockton et al. 2000
Otospondylomega-epiphyseal dysplasia	Col IIA2	Cleft palate Flat face Microagnathia Upturned nose Hearing loss	Short limbs Large skeletal epiphysis	Missense or splicing mutations	215150 120290	Melkoniemi et al. 2000
Treacher-Collins	TREACLE (nuclear phosphoprotein)	Mandible and maxillary hypoplasia Cleft palate Ear abnormalities Eyelid defects		haploinsufficiency	154500 606847	Treacher-Collins Collaborative Group (see Sect. 8.2)
Tooth agenesis and clefting	Msx1	lack of mandibular and maxillary premolars third molar frequently absent Cleft lip and/or palate		Loss-of-function Defect in epithelial-mesenchymal interactions	106600 142983	Vastardis et al. 1996 Hu et al. 1998 van den Boogaard et al. 2000
Skeletal defects						
Apert	FGFR2	Craniosynostosis	Hand and foot defects	Gain-of-function	101200 176943	Morriss-Kay et al. 2001 Wilkie et al. 2001
Boston-type craniosynostosis	MSX2	Craniosynostosis		Gain-of-function	604757 123101	Morriss-Kay et al. 2001 Wilkie et al. 2001
Crouzon	FGFR2 and 3	Craniosynostosis	Skin defects	Gain-of-function	123500 176943 134934	Morriss-Kay et al. 2001 Wilkie et al. 2001

Table 5. *Continued*[a]

Syndrome	Gene	Phenotype: Craniofacial	Phenotype: Other	Mechanism of mutation	Omim reference	References
Jackson-Weiss1	FGFR2	Craniosynostosis	Foot defects	Gain-of-function	123150	Morriss-Kay et al. 2001 Wilkie et al. 2001
Pfeiffer	FGFR1 and 2	Craniosynostosis	Hand and foot defects	Gain-of-function	101600 136350	Morriss-Kay et al. 2001 Wilkie et al. 2001
Saethre-Chotzen	TWIST Transcription factor	Craniosynostosis (coronal suture) Low frontal hair line and high forehead Facial asymmetry Nasal defects Hypertelorism Maxillary hypoplasia Cleft palate	Limb defects	Haploinsufficiency Mutations decrease protein stability or prevent heterodimer formation or prevent DNA binding	101400 601622	El Ghouzzi et al. 1997 El Chouzzi et al. 1999 El Ghouzzi et al. 2001 Howard et al. 1997
Enlarged parietal foramina (PFM)	MSX2	Oval defects in parietal bones	Brain defects Scalp defects	Loss-of-function	168500 123101	Wilkie et al. 2000
	ALX4	Oval defects in parietal bones Cranium bifidum		Loss-of-function	168500 601224 605420	Mavrogiannis et al. 2001
Craniometaphyseal dysplasia1	ANKH (membrane protein)	Craniofacial bones are thicker	Long bones are thicker	Gain-of-function	123000 605145	Nürnberg et al. 2001
Tricho-dento-osseous syndrome-2	DLX3	Dental defects – taurodontism and enamel hypoplasia Skeletal defects – cranial bones are thicker		Deletion mutant	190320 600525	Price et al. 1998
Faciogenital dysplasia (Aarskog-Scott) X-linked	FGD1 (Rho GTPase activator)	Hypertelorism Upturned and short nose	Short stature Mental retardation Hand defects Joint defects Genital defects	Not known	305400	Orrico et al. 2000 Pasteris et al. 1994 Gorski et al. 2000

Table 5. *Continued*[a]

Syndrome	Gene	Phenotype		Mechanism of mutation	Omim reference	References
		Craniofacial	Other			
Tricho-rhino-phalangeal syndrome type 1	TRPS 1 (transcription factor)	Nose abnormalities Protruding ears	Short stature Hip malformations Digit malformations Sparse scalp hair	Haploinsufficiency	190350 604386	Hatamura et al. 2001 Momeni et al. 2000

[a]The syndromes are divided into: "Midline defects," "Exencephaly," "Facial defects," and "Skeletal defects." The phenotypes of each syndrome are divided into craniofacial and "other" defects. The type of mutation and/or how it acts are reported under the heading "Mechanism." The Omim Number refers to the Online Mendelian Inheritance in Man (www.ncbi.nlm.nih.gov/omim) references which describe in more detail the phenotypes and the genes responsible. Please note that other genes which have yet to be identified may also be responsible for these syndromes.

and Camper 1997; Arakawa et al. 1998; Mitsiadis et al. 1998; Hjalt et al. 2000; 2001). In addition, *Pitx2* is found asymmetrically in the early lateral plate mesoderm where it plays a role in the left-right patterning of the body axis (Logan et al. 1998; Ryan et al. 1998; Yoshioka et al. 1998; Gage et al. 1999; Lin et al. 1999). Gene inactivation studies in mice and the generation of hypomorphs have shown that various tissues are differentially sensitive to loss of *Pitx2* (Lin et al. 1999; Gage et al. 1999). For example, the eyes and the heart are more susceptible than the pituitary, which may in part reflect functional redundancy as *Pitx1* is co-expressed with *Pitx2* in the latter. Analysis of the tooth defects in homozygous mutant mice has revealed that tooth development is initiated correctly but does not progress beyond the placode stage in the maxillary teeth or the tooth bud stage in the mandibular primordia (Lin et al. 1999; Lu et al. 1999b).

The mutations identified to date include missense mutations within the homeodomain and splicing mutations. Patients with a deletion in the *PITX2* gene or with translocation break points upstream of the *PITX2* gene, which presumably affect the levels of *PITX2* transcripts, have also been identified (Datson et al. 1996; Semina et al. 1996; Flomen et al. 1998). Whilst all the mutations identified within the homeobox domain abolish activation of the pituitary-specific prolactin promoter, which contains multiple bicoid target sites, there are distinct differences in their mechanism which may explain the variable phenotypes seen in Riegers syndrome. The mutation T68P reduces DNA binding and prevents transactivation of reporter constructs whilst the L54Q mutation appears to destabilise the Pitx2 protein, resulting in its degradation (Amendt al. 1998; Kozlowski and Walter 2000; Saadi et al. 2001). The K88E mutation converts a lysine to glutamic acid at position 50, which is predicted to interact with the DNA consensus-binding site. This mutant also does not bind to the bicoid target site. However, in contrast to T68P, the K88E mutant protein suppresses the ability of Pitx2/Pit1 to synergistically transactivate DNA target sites (Saadi et al. 2001). Hence, this mutant protein acts as a dominant-negative. Other mutations result in truncation of the Pitx2 protein, resulting in the loss of the C terminal domain. This domain represses DNA binding activity, possibly by binding to the N terminal domain, making the homeobox domain inaccessible for DNA binding. The C terminus is also required for Pitx2-Pit1 interaction, and it has been proposed that Pit-1 binding to the C terminus releases its inhibitory action, making the homeodomain available for DNA binding (Amendt et al. 1999). This would be one mechanism for their synergistic interactions. Two candidate down-stream targets of Pitx2 have been identified – procollagen lysyl hydrolyase 1 and 2 (PLOD 1 and 2) (Hjalt et al. 2001). These proteins hydrolyse lysines in collagens. Mutation of the *PLOD1* in humans results in Ehlers-Danlos syndrome, kyphoscoliosis-type (type VI; EDVI), which has overlapping phenotypes with Riegers syndrome such as the ocular (microcornea, glaucoma, cornea plana) and abdominal (inguinal hernia, umbilical hernia) abnormalities (Hyland et al. 1992; Heikkinen et al. 1997). In contrast, the dental abnormalities are distinct, with EDVI patients exhibiting dentinogenesis imperfecta.

8.2 Treacher-Collins Syndrome

This autosomal dominant syndrome is the most common mandibulofacial dysostosis, affecting between 1 in 25, 000 to 1 in 50, 000 births with 60% of cases arising as new mutations (Jones et al. 1975). The phenotypic consequences are very heterogeneous but are predominantly characterised by first and second branchial arch defects which result in hypoplastic mid- and lower facial skeletal structures and hence, micrognathia. Due to hypoplasia of the malar (cheek) bones, the midface is convex. There is conductive hearing loss due to the absence or reduction in the external and middle ears (microtia), although the inner ear is normal. The palpebral fissures slant downwards, and there is colobomata of the lower eyelids with loss of the lower eyelashes medial to this defect. There may be cleft lip with or without cleft palate. In addition, the neurocranium may be abnormally shaped (reviewed by Posnick and Ruiz 2000).

Treacher-Collins syndrome is due to mutations in the *TCOF1* gene, which encodes treacle (Gladwin et al. 1996; Treacher-Collins Group 1996; Edwards et al. 1997). Treacle is a low-complexity protein comprising three domains: a unique C and N terminus separated by a highly phosphorylated central domain containing ten serine and acidic amino-acid repeats which are target phosphorylation sites for casein kinase II (CK2) and protein kinase C (PKC) (Gladwin et al. 1996; Treacher-Collins Group 1996; Dixon et al. 1997a,b; Edwards et al. 1997; Paznekas et al. 1997; Wise et al. 1997; Isaac et al. 2000). There are also nuclear localisation signals in the C terminus (Dixon et al. 1997b; Wise et al. 1997; Marsh et al. 1998; Winokur and Shiang 1998). Over 50 distinct mutations that result in Treacher-Collins syndrome have been identified, the majority of which result in protein truncations and hence loss of the nuclear signal sequence. Consistent with this, antibody immunostaining has localised treacle, but not mutated forms of treacle, to the nucleolus (Marsh et al. 1998; Winokur and Shiang 1998; Isaac et al. 2000).

The central repeat domain of treacle shares homology, primarily at the predicted PKC and CK2 phosphorylation sites, with the nuclear phosphorylation protein Nopp140, which when phosphorylated shuttles along fibrillar tracks between the nucleolus and cytoplasm, acting as a protein chaperone (Meier and Blobel 1992; Cairns and McStay 1995; Dixon et al. 1997b; Wise et al. 1997). Nopp140 has also been shown to act as a transcriptional co-activator (Miau et al. 1997). The exact role of treacle is unknown. However, like Nopp140, treacle is also associated with the fribillin network and may play a role in nucleolar-cytoplasmic shuttling (Isaac et al. 2000).

In situ expression analysis in mice has shown that *treacle* is expressed ubiquitously with particularly high levels of expression in the neural folds prior to fusion and later in the face (Dixon et al. 1997b). Activity of casein kinase II-like molecules, which can phosphorylate treacle, has been detected at higher levels in the branchial arches (Jones et al. 1999). Gene inactivation studies in mice have shown that heterozygosity of treacle results in overlapping phenotypes with Treacher-Collins syndrome, although the phenotypes tend to be more severe. First, as in humans, there is hypoplasticity of the branchial arches, with malformations of the nasal pits and maxilla. The otocyst is also smaller (Dixon et al. 2000). However, in contrast to humans, exencephaly, anophthalmia, and upper frontonasal mass/skull defects are observed. *Treacle* heterozygote mice are also much smaller. Further analysis has revealed that in *treacle* heterozygotes, there is increased apoptosis in the neuroepithelium and head region at E9. Treacle may also play a more general role in growth, possibly by modulating ribosome synthesis and assembly.

The Treacher-Collins syndrome has been proposed to result from either a dominant-negative or haploinsufficiency effect. Current evidence favours the latter. First, gene inactivation studies in mice have shown that haploinsufficiency of *treacle* results in craniofacial defects (Dixon et al. 2000). Second, as a result of the mutations the truncated protein products appear to be rapidly degraded (Marsh et al. 1998; Winokur and Shiang 1998; Isaac et al. 2000). Third, disease-causing mutations are found throughout the gene. Furthermore, overexpression of these truncated proteins *in vitro* does not block translocation of full-length treacle into the nucleus (Isaac et al. 2000). Paradoxically, the levels of wild-type treacle in cells from treacle patients are the same as in unaffected individuals, suggesting that the levels of the wild-type protein may be up-regulated to compensate for the loss of one functional allele (Isaac et al. 2000). However, the latter analysis was carried out in fibroblasts and lymphoblasts, and as such may not be representative as to what occurs in other cell types such as the neural crest. Although ubiquitously expressed, the craniofacial structures are specifically affected. The reason for this is unclear. However, this might reflect the higher-level expression of treacle, and growth in the craniofacial region, or alternatively, treacle may be critical for the function, for example by acting as a chaperone or as a transcriptional co-factor, of a protein that is specifically expressed in the craniofacial region.

8.3 DiGeorge Syndrome

This haploinsufficiency syndrome is associated with deletion of part of chromosome 22, region q11.21–q11.23, in over 90% of the cases and shares many common features with velocardiofacial syndrome (VCFS), also frequently due to deletions of this chromosome region. Another locus that has been implicated in DiGeorge syndrome (DGS) is 10p13–p14 (Daw et al. 1996; Schuffenhauer et al. 1998). Together with conotruncal-anomaly face and Opitz syndrome, these are termed the CATCH22 syndromes. DGS occurs in 1 in 4,000 births and is the most common genetic cause of both heart and facial birth defects. There is slight facial dysmorphism including micrognathia, low-set ears with abnormally folded helices, and cleft palate. The major defects are heart blood vessel abnormalities, which most commonly are interruptions of the aortic arch, persistent truncus arteriosus, and misalignment of the aorta and pulmonary arteries. The latter is also known as tetralogy of Fallot. In addition, there is hypoparathyroidism, resulting in hypocalcemia, and hypoplasticity/aplasia of the thymus resulting in immunodeficiency (Wilson et al. 1993). There may also be behavioural and learning difficulties. The majority of the phenotypes of DiGeorge syndrome resemble those following neural crest ablation, and it has been proposed that abnormal development of the cranial neural crest is responsible (Lammer and Opitz 1986; Kirby and Waldo 1990; Goldmuntz and Emanuel 1997). However, recent evidence has suggested that the cause may not be this simple and that the development of other tissues, which control neural crest survival and/or proliferation, may be affected. For example, *TBX1*, which has been heavily implicated in DGS, is not expressed in the neural crest cells or its derivatives (Chapman et al. 1996; Jerome and Papaionnou 2001; Merscher et al. 2001).

The approaches taken to identify DiGeorge candidates include gene mapping in humans to determine which genes are deleted and/or mutated, expression analysis of candidate genes in developing embryos, together with loss of function studies in mice or chicks (reviewed by Goldmuntz and Emanuel 1997). However, the situation may be more complex than just a gene deletion. In some DGS patients, the deletions do not overlap whilst in others there is no deletion of chromosome 22 – yet in these cases, no mutations in the genes in this region have been identified. Therefore, long-range effects due to conformational changes in the chromatin structure or the loss of regulatory elements may act on genes outside the critical region. In addition, to further complicate matters, two minimum critical regions have been identified (Levy et al. 1995; Gong et al. 1996; Kurahashi et al. 1997).

The generation of four mouse models with deletions of parts of the syntenic region on chromosome 16 has also significantly advanced the research (Kimber et al. 1999; Lindsay et al. 1999; Puech et al. 2000; Merscher et al. 2001). Mice with a 1.2-Mb deletion, *Df1*, have cardiac defects in a portion of the heterozygotes and this, in part, ultimately helped to pin down the T-box transcription factor, *Tbx1*, as possibly being the key player responsible for the cardiac defects in humans (Lindsay et al. 1999; Jerome and Papaionnou 2001; Merscher et al. 2001). In contrast, another mouse mutant with a 140-kb deletion, encompassing seven genes and overlapping with the *Df1* deletion, has behavioural abnormalities, showing that this region is essential for normal neural development (Kimber et al. 1999).

22q11 contains 24 genes including *UFD1* (ubiquitin fusion degradation 1), *HIRA* (also called *TUPLE1*), *CRKL*, and the transcription factors *GSC1*, *TBX1*, *ZNF74*, and *LZTR-1*. Of these, *UFD1*, *HIRA*, *CRKL*, and *TBX1* have been heavily implicated as the causative gene responsible for at least a subset of abnormalities. In all cases, however, a single gene has not been shown to be either mutated or deleted in every DiGeorge patient, suggesting that there is not one master gene that will be identified. In addition, there is extreme heterogeneity among DiGeorge patients, with even monozygotic twins exhibiting different dysmorphic features. Furthermore, the mouse model used as an experimental system does not always equate directly with humans. For example, deletion of *Pax3* in mice results in the *Splotch* mouse characterised by outflow tract together with thymus and parathyroid abnormalities (Conway et al. 1997a,b). In contrast, mutation of *PAX3*, although not necessarily equivalent to the mouse deletion, in humans results in Waardenburg syndrome type 1, characterised by hearing loss and hypopigmentation (Hoth et al. 1993; Tassabehji et al. 1993; Baldwin et al. 1994, 1995).

Attention initially focused on *UFD1* and *CDC45L2*, as these were the only genes deleted in one DiGeorge patient (Yamagishi et al. 1999). These genes are outside the 250-Kb minimum DiGeorge critical region, but are deleted in the majority of DGS and VCFS patients (Gong et al. 1996). Cdc45l, which is required for the initiation of DNA replication, and *Ufd1*, a protein implicated in the ubiquitin-dependent proteolytic pathway, are expressed in the branchial arches and cardiac outflow tract (Yamagishi et al. 1999; Kunte et al. 2001). Based on the role of its homologous protein in yeast, Ufd1 has also been suggested to regulate RNA polyadenylation. Ubiquitin-protein conjugates are abundant in areas of remodelling such as during formation of the aortic arches, suggesting that Ufd1 may play a role in their formation (Wunsch and Haas 1995). Expression of *Ufd1*, but not *cdc45l*, is down-regulated in *dHAND* (also known as hand2) mouse mutants, which have similar cardiac defects, in part due to increased apoptosis, to DiGeorge patients (Yamagishi et al. 1999). This suggests that *Ufd1* is downstream of dHAND and that the cardiac phenotypes in the *dHAND* mutants may be due to loss of *Ufd1* expression (Thomas et al. 1998; Srivastava et al. 1997). This has led to the proposal that in the absence of UFD1, protein accumulation results in excessive apoptosis of the neural crest cells. However, an argument against this protein being a major player in cardiac development is that gene inactivation of *Ufd1* in mice has no phenotypic consequences (Lindsay et al. 1999). Furthermore, DGS patients who do not have deletions or mutations in the *UFD1* gene have been identified (Wadey et al. 1999).

Another candidate gene is *HIRA*, which shares homology to the yeast genes *hir1p* and *hir2p*. These yeast proteins repress histone gene transcription and have been proposed to remodel chromatin (Lamour et al. 1995; Spector et al. 1997). HIRA is also homologous to Tup-1, a yeast protein, which interacts with histones and transcription factors. Protein binding studies have shown that HIRA binds histones and other proteins including PAX3, which in mice is required for conotruncal septation and thymus/parathyroid development (Conway et al. 1997a,b; Roberts et al. 1997; Lorain et al. 1998; Magnaghi et al. 1998). Therefore, like the yeast homologues, HIRA may act as a transcriptional cofactor and alter chromatin structure modulating transcription. *Hira* is expressed at elevated levels in the neural folds, neural crest, and its derivatives (Roberts et al. 1997; Wilming et al. 1997). In the chick, ablation of *Hira* mRNA in the cardiac neural crest with antisense oligos specifically results in persistent truncus arteriosus but does not affect aortic arch patterning (Farrell et al. 1999).

Recent work has suggested that the transcription factor TBX1 is likely to be the single causative gene for the heart defects. TBX1 is related to TBX5, which when mutated in humans causes Holt-Oram syndrome, also characterised by cardiac defects (Basson et al. 1997, 1999; Li et al. 1997b). Heterozygosity of *Tbx1* in mice results in specific defects in the fourth arch arteries, whilst homozygotes have additional cardiac defects that encompass all those found in DGS patients (Jerome and Papaioannou 2001; Lindsay et al. 2001; Merscher et al. 2001). *Tbx1* is expressed in the mesenchyme underlying the hindbrain, pharyngeal arches and pouches, but not the neural crest-derived mesenchyme (Chapman et al. 1996; Lindsay et al. 2001; Merscher et al. 2001). This suggests that Tbx1 may regulate the expression of factors that control neural crest survival and/or proliferation. As there are early segmentation defects in the heterozygous mice, it has been proposed that Tbx1 is required for endoderm-mesenchyme interactions during formation of the arches (Lindsay et al. 2001). The reduced dosage-sensitivity in mice, and hence presentation of the majority of the "DGS" phenotypes in homozygote but not heterozygote mice, may simply reflect species differences between mouse and humans, as has been seen with the *MSX2*, *PAX9*, and *PAX3* genes (see Tables 3 and 5).

Finally, the *CRKL* gene, which encodes an SH2-SH3-SH3 adaptor protein thought to be involved in growth factor and focal adhesion signalling, has also been implicated in the pathogenesis of DiGeorge syndrome. Again the gene is in the DiGeorge deletion region, and it is expressed in the defective tissues – the neuroepithelium, neural crest, and the branchial arches (Guris et al. 2001). Homozygous but not heterozygote mice exhibit outflow tract, aortic arch, thymus, and parathyroid abnormalities together with mild deformities in the skull, again consistent with an essential role in the target tissues affected by 22p11 deletions. There are also defects in the cranial nerves, which have not been reported in DGS patients. Analysis of neural crest development in the

Crkl null mutants has shown that crkl is not required for neural crest generation or survival, suggesting that it is required later during craniofacial morphogenesis and differentiation. Indeed, the pharyngeal arch arteries of the *Crkl* mutants initially form correctly but fail to undergo the later steps of remodelling (Guris et al. 2001).

Summary

The molecular cascades that control craniofacial development have until recently been little understood. The paucity of data that exists has in part been due to the complexity of the head, which is the most intricate regions of the body. However, the generation of mouse mutants and the identification of gene mutations that cause human craniofacial syndromes, together with classical embryological approaches in other species, have given significant insight into how the head develops. These studies have emphasized how unique the head actually is, with each individual part governed by a distinct set of signalling interactions, again demonstrating the complexity of this region of the body. This review discussed the tissue and molecular interactions that control each region of the head. The processes that control neural tube closure together with correct development of the skull, midline patterning, neural crest generation and migration, outgrowth, patterning, and differentiation of the facial primordia and the branchial arches are thus discussed. Defects in these processes result in a number of human syndromes such as exencephaly, holoprosencephaly, musculoskeletal dysplasias, first arch syndromes such as Riegers and Treacher-Collins syndrome, and neural crest dysplasias such as DiGeorge syndrome. Our current knowledge of the genes responsible for these human syndromes together with how the head develops, is rapidly advancing so that we will soon understand the complex set of molecular and tissue interactions that build a head.

References

Abbott, B.D. (1995): Review of the interaction between TCDD and glucocorticoids in embryonic palate. Toxicology, 105: 365–373

Abbott, B.D. and Pratt, R.M. (1987a): Human embryonic palatal epithelial differentiation is altered by retinoic acid and epidermal growth factor in organ culture. J Craniofac Genet Dev Biol, 7: 241–265

Abbott, B.D. and Pratt, R.M. (1987b): Retinoids and epidermal growth factor alter embryonic mouse palatal epithelial and mesenchymal cell differentiation in organ culture. J Craniofac Genet Dev Biol, 7: 219–240

Abbott, B.D. and Pratt, R.M. (1991): Retinoic acid alters epithelial differentiation during palatogenesis. J Craniofac Genet Dev Biol, 11: 315–325

Acampora, D., Merlo, G.R., Paleari, L., Zerega, B., Postiglione, M.P., Mantero, S., Bober, E., Barbieri, O., Simeone, A. and Levi, G. (1999): Craniofacial, vestibular and bone defects in mice lacking the Distal- less-related gene Dlx5. Development, 126: 3795–3809

Adams, R.H., Wilkinson, G.A., Weiss, C., Diella, F., Gale, N.W., Deutsch, U., Risau, W. and Klein, R. (1999): Roles of ephrinB ligands and EphB receptors in cardiovascular development: demarcation of arterial/venous domains, vascular morphogenesis, and sprouting angiogenesis. Genes Dev, 13: 295–306

Adams, R.H., Diella, F., Hennig, S., Helmbacher, F., Deutsch, U. and Klein, R. (2001): The cytoplasmic domain of the ligand ephrinB2 is required for vascular morphogenesis but not cranial neural crest migration. Cell, 104: 57–69

Agius, E., Oelgeschlager, M., Wessely, O., Kemp, C. and De Robertis, E.M. (2000): Endodermal Nodal-related signals and mesoderm induction in Xenopus. Development, 127: 1173–1183

Ahlgren, S.C. and Bronner-Fraser, M. (1999): Inhibition of sonic hedgehog signaling in vivo results in craniofacial neural crest cell death. Curr Biol, 9: 1304–1314

Akazawa, H., Komuro, I., Sugitani, Y., Yazaki, Y., Nagai, R. and Noda, T. (2000): Targeted disruption of the homeobox transcription factor Bapx1 results in lethal skeletal dysplasia with asplenia and gastroduodenal malformation. Genes Cells, 5: 499–513

Akitaya, T. and Bronner-Fraser, M. (1992): Expression of cell adhesion molecules during initiation and cessation of neural crest cell migration. Dev Dyn, 194: 12–20

Alfandari, D., Cousin, H., Gaultier, A., Smith, K., White, J.M., Darribere, T. and DeSimone, D.W. (2001): Xenopus ADAM 13 is a metalloprotease required for cranial neural crest- cell migration. Curr Biol, 11: 918–930

Amendt, B.A., Sutherland, L.B., Semina, E.V. and Russo, A.F. (1998): The molecular basis of Rieger syndrome. Analysis of Pitx2 homeodomain protein activities. J Biol Chem, 273: 20066–20072

Amendt, B.A., Sutherland, L.B. and Russo, A.F. (1999): Multifunctional role of the Pitx2 homeodomain protein C-terminal tail. Mol Cell Biol, 19: 7001–7010

Amendt, B.A., Semina, E.V. and Alward, W.L. (2000): Rieger syndrome: a clinical, molecular, and biochemical analysis. Cell Mol Life Sci, 57: 1652–1666

Anderson, C.B. and Meier, S. (1981): The influence of the metameric pattern in the mesoderm on migration of cranial neural crest cells in the chick embryo. Dev Biol, 85: 385–402

Arakawa, H., Nakamura, T., Zhadanov, A.B., Fidanza, V., Yano, T., Bullrich, F., Shimizu, M., Blechman, J., Mazo, A., Canaani, E. and Croce, C.M. (1998): Identification and characterization of the ARP1 gene, a target for the human acute leukemia ALL1 gene. Proc Natl Acad Sci U S A, 95: 4573–4578

Arikawa-Hirasawa, E., Watanabe, H., Takami, H., Hassell, J.R. and Yamada, Y. (1999): Perlecan is essential for cartilage and cephalic development. Nat Genet, 23: 354–358.

Arkell, R. and Beddington, R.S. (1997): BMP-7 influences pattern and growth of the developing hindbrain of mouse embryos. Development, 124: 1–12

Aruga, J., Mizugishi, K., Koseki, H., Imai, K., Balling, R., Noda, T. and Mikoshiba, K. (1999): Zic1 regulates the patterning of vertebral arches in cooperation with Gli3. Mech Dev, 89: 141–150.

Asada, H., Kawamura, Y., Maruyama, K., Kume, H., Ding, R.G., Kanbara, N., Kuzume, H., Sanbo, M., Yagi, T. and Obata, K. (1997): Cleft palate and decreased brain gamma-aminobutyric acid in mice lacking the 67-kDa isoform of glutamic acid decarboxylase. Proc Natl Acad Sci U S A, 94: 6496–6499

Bachiller, D., Klingensmith, J., Kemp, C., Belo, J.A., Anderson, R.M., May, S.R., McMahon, J.A., McMahon, A.P., Harland, R.M., Rossant, J. and De Robertis, E.M. (2000): The organizer factors Chordin and Noggin are required for mouse forebrain development. Nature, 403: 658–661

Bachler, M. and Neubüser, A. (2001): Expression of members of the Fgf family and their receptors during midfacial development. Mech Dev, 100: 313–316

Bader, B.L., Rayburn, H., Crowley, D. and Hynes, R.O. (1998): Extensive vasculogenesis, angiogenesis, and organogenesis precede lethality in mice lacking all alpha v integrins. Cell, 95: 507–519

Baker, C.V. and Bronner-Fraser, M. (2001): Vertebrate cranial placodes I. Embryonic induction. Dev Biol, 232: 1–61

Baldwin, C.T., Lipsky, N.R., Hoth, C.F., Cohen, T., Mamuya, W. and Milunsky, A. (1994): Mutations in PAX3 associated with Waardenburg syndrome type I. Hum Mutat, 3: 205–211

Baldwin, C.T., Hoth, C.F., Macina, R.A. and Milunsky, A. (1995): Mutations in PAX3 that cause Waardenburg syndrome type I: ten new mutations and review of the literature. Am J Med Genet, 58: 115–122

Bang, A.G., Papalopulu, N., Kintner, C. and Goulding, M.D. (1997): Expression of Pax-3 is initiated in the early neural plate by posteriorizing signals produced by the organizer and by posterior non-axial mesoderm. Development, 124: 2075–2085

Bardot, B., Lecoin, L., Huillard, E., Calothy, G. and Marx, M. (2001): Expression pattern of the drm/gremlin gene during chicken embryonic development. Mech Dev, 101: 263–265

Barembaum, M., Moreno, T.A., LaBonne, C., Sechrist, J. and Bronner-Fraser, M. (2000): Noelin-1 is a secreted glycoprotein involved in generation of the neural crest. Nat Cell Biol, 2: 219–225

Barlow, A.J. and Francis-West, P.H. (1997): Ectopic application of recombinant BMP-2 and BMP-4 can change patterning of developing chick facial primordia. Development, 124: 391–398

Barlow, A.J., Bogardi, J.P., Ladher, R. and Francis-West, P.H. (1999): Expression of chick Barx-1 and its differential regulation by FGF-8 and BMP signaling in the maxillary primordia. Dev Dyn, 214: 291–302

Barlow, L.A. (2001): Specification of pharyngeal endoderm is dependent on early signals from axial mesoderm. Development, 128: 4573–4583

Barni, T., Maggi, M., Fantoni, G., Serio, M., Tollaro, I., Gloria, L. and Vannelli, G.B. (1995): Identification and localization of endothelin-1 and its receptors in human fetal jaws. Dev Biol, 169: 373–377

del Barrio, M.G. and Nieto, M.A. (2002): Overexpression of snail family members highlights their ability to promote chick neural crest formation. Development, 129: 1583–1593

Barrow, J.R. and Capecchi, M.R. (1999): Compensatory defects associated with mutations in Hoxa1 restore normal palatogenesis to Hoxa2 mutants. Development, 126: 5011–5026

Barth, K.A. and Wilson, S.W. (1995): Expression of zebrafish nk2.2 is influenced by sonic hedgehog/vertebrate hedgehog-1 and demarcates a zone of neuronal differentiation in the embryonic forebrain. Development, 121: 1755–1768

Basch, M.L., Selleck, M.A. and Bronner-Fraser, M. (2000): Timing and competence of neural crest formation. Dev Neurosci, 22: 217–227

Basson, C.T., Bachinsky, D.R., Lin, R.C., Levi, T., Elkins, J.A., Soults, J., Grayzel, D., Kroumpouzou, E., Traill, T.A., Leblanc-Straceski, J., Renault, B., Kucherlapati, R., Seidman, J.G. and Seidman, C.E. (1997): Mutations in human TBX5 cause limb and cardiac malformation in Holt-Oram syndrome. Nat Genet, 15: 30–35

Battaile, K.P. and Steiner, R.D. (2000): Smith-Lemli-Opitz syndrome: the first malformation syndrome associated with defective cholesterol synthesis. Mol Genet Metab, 71: 154–162

Baynash, A.G., Hosoda, K., Giaid, A., Richardson, J.A., Emoto, N., Hammer, R.E. and Yanagisawa, M. (1994): Interaction of endothelin-3 with endothelin-B receptor is essential for development of epidermal melanocytes and enteric neurons. Cell, 79: 1277–1285

Beall, A.C. and Rosenquist, T.H. (1990): Smooth muscle cells of neural crest origin form the aorticopulmonary septum in the avian embryo. Anat Rec, 226: 360–366

Beddington, R.S. (1994): Induction of a second neural axis by the mouse node. Development, 120: 613–620

Bee, J. and Thorogood, P. (1980): The role of tissue interactions in the skeletogenic differentiation of avian neural crest cells. Dev Biol, 78: 47–62

Bei, M. and Maas, R. (1998): FGFs and BMP4 induce both Msx1-independent and Msx1-dependent signaling pathways in early tooth development. Development, 125: 4325–4333

Belloni, E., Muenke, M., Roessler, E., Traverso, G., Siegel-Bartelt, J., Frumkin, A., Mitchell, H.F., Donis-Keller, H., Helms, C., Hing, A.V., Heng, H.H., Koop, B., Martindale, D., Rommens, J.M., Tsui, L.C. and Scherer, S.W. (1996): Identification of Sonic hedgehog as a candidate gene responsible for holoprosencephaly. Nat Genet, 14: 353–356

Belo, J.A., Leyns, L., Yamada, G. and De Robertis, E.M. (1998): The prechordal midline of the chondrocranium is defective in Goosecoid- 1 mouse mutants. Mech Dev, 72: 15–25.

Bennett, J.H., Hunt, P. and Thorogood, P. (1995): Bone morphogenetic protein-2 and –4 expression during murine orofacial development. Arch Oral Biol, 40: 847–854

Berk, M., Desai, S.Y., Heyman, H.C. and Colmenares, C. (1997): Mice lacking the ski proto-oncogene have defects in neurulation, craniofacial, patterning, and skeletal muscle development. Genes Dev, 11: 2029–2039

Bertolino, E., Reimund, B., Wildt-Perinic, D. and Clerc, R.G. (1995): A novel homeobox protein which recognizes a TGT core and functionally interferes with a retinoid-responsive motif. J Biol Chem, 270: 31178–31188

Beverdam, A., Brouwer, A., Reijnen, M., Korving, J. and Meijlink, F. (2001): Severe nasal clefting and abnormal embryonic apoptosis in Alx3/Alx4 double mutant mice. Development, 128: 3975–3986

Birgbauer, E. and Fraser, S.E. (1994): Violation of cell lineage restriction compartments in the chick hindbrain. Development, 120: 1347–1356

Birgbauer, E., Sechrist, J., Bronner-Fraser, M. and Fraser, S. (1995): Rhombomeric origin and rostrocaudal reassortment of neural crest cells revealed by intravital microscopy. Development, 121: 935–945

Bisgrove, B.W., Essner, J.J. and Yost, H.J. (1999): Regulation of midline development by antagonism of lefty and nodal signaling. Development, 126: 3253–3262

Blader, P. and Strähle, U. (1998): Ethanol impairs migration of the prechordal plate in the zebrafish embryo. Dev Biol, 201: 185–201

Bladt, F., Riethmacher, D., Isenmann, S., Aguzzi, A. and Birchmeier, C. (1995): Essential role for the c-met receptor in the migration of myogenic precursor cells into the limb bud. Nature, 376: 768–771

Bockman, D.E. and Kirby, M.L. (1984): Dependence of thymus development on derivatives of the neural crest. Science, 223: 498–500

Bockman, D.E. and Kirby, M.L. (1985): Neural crest interactions in the development of the immune system. J Immunol, 135: 766s-768s

Bockman, D.E., Redmond, M.E., Waldo, K., Davis, H. and Kirby, M.L. (1987): Effect of neural crest ablation on development of the heart and arch arteries in the chick. Am J Anat, 180: 332–341

Bockman, D.E., Redmond, M.E. and Kirby, M.L. (1989): Alteration of early vascular development after ablation of cranial neural crest. Anat Rec, 225: 209–217

Bockman, D.E. and Sohal, G.S. (1998): A new source of cells contributing to the developing gastrointestinal tract demonstrated in chick embryos. Gastroenterology, 114: 878–882

Bonstein, L., Elias, S. and Frank, D. (1998): Paraxial-fated mesoderm is required for neural crest induction in Xenopus embryos. Dev Biol, 193: 156–168

Borchers, A., David, R. and Wedlich, D. (2001): Xenopus cadherin-11 restrains cranial neural crest migration and influences neural crest specification. Development, 128: 3049–3060

Borello, U., Coletta, M., Tajbakhsh, S., Leyns, L., De Robertis, E.M., Buckingham, M. and Cossu, G. (1999): Transplacental delivery of the Wnt antagonist Frzb1 inhibits development of caudal paraxial mesoderm and skeletal myogenesis in mouse embryos. Development, 126: 4247–4255

Boucaut, J.C., Darribere, T., Poole, T.J., Aoyama, H., Yamada, K.M. and Thiery, J.P. (1984): Biologically active synthetic peptides as probes of embryonic development: a competitive peptide inhibitor of fibronectin function inhibits gastrulation in amphibian embryos and neural crest cell migration in avian embryos. J Cell Biol, 99: 1822–1830
Brais, B., Bouchard, J.P., Xie, Y.G., Rochefort, D.L., Chretien, N., Tome, F.M., Lafreniere, R.G., Rommens, J.M., Uyama, E., Nohira, O., Blumen, S., Korczyn, A.D., Heutink, P., Mathieu, J., Duranceau, A., Codere, F., Fardeau, M., Rouleau, G.A., Korcyn, A.D. (1998): Short GCG expansions in the PABP2 gene cause oculopharyngeal muscular dystrophy. Nat Genet, 18:164–7
Brand, M., Heisenberg, C.P., Warga, R.M., Pelegri, F., Karlstrom, R.O., Beuchle, D., Picker, A., Jiang, Y.J., Furutani-Seiki, M., van Eeden, F.J., Granato, M., Haffter, P., Hammerschmidt, M., Kane, D.A., Kelsh, R.N., Mullins, M.C., Odenthal, J., Nusslein-Volhard, C. (1996): Mutations affecting development of the midline and general body shape during zebrafish embryogenesis. Development, 123:129–42
Brand, M., Le Moullec, J.M., Corvol, P. and Gasc, J.M. (1998): Ontogeny of endothelins-1 and -3, their receptors, and endothelin converting enzyme-1 in the early human embryo. J Clin Invest, 101: 549–559
Brault, V., Moore, R., Kutsch, S., Ishibashi, M., Rowitch, D.H., McMahon, A.P., Sommer, L., Boussadia, O. and Kemler, R. (2001): Inactivation of the beta-catenin gene by Wnt1-Cre-mediated deletion results in dramatic brain malformation and failure of craniofacial development. Development, 128: 1253–1264
Brewster, R., Lee, J. and Ruiz i Altaba, A. (1998): Gli/Zic factors pattern the neural plate by defining domains of cell differentiation. Nature, 393: 579–583
Brohmann, H., Jagla, K. and Birchmeier, C. (2000): The role of Lbx1 in migration of muscle precursor cells. Development, 127: 437–445
Bronner-Fraser, M. (1985): Alterations in neural crest migration by a monoclonal antibody that affects cell adhesion. J Cell Biol, 101: 610–617
Bronner-Fraser, M. (1986): An antibody to a receptor for fibronectin and laminin perturbs cranial neural crest development in vivo. Dev Biol, 117: 528–536
Bronner-Fraser, M. (1987): Perturbation of cranial neural crest migration by the HNK-1 antibody. Dev Biol, 123: 321–331
Bronner-Fraser, M. (1988): Distribution and function of tenascin during cranial neural crest development in the chick. J Neurosci Res, 21: 135–147
Bronner-Fraser, M. and Lallier, T. (1988): A monoclonal antibody against a laminin-heparan sulfate proteoglycan complex perturbs cranial neural crest migration in vivo. J Cell Biol, 106: 1321–1329
Bronner-Fraser, M., Wolf, J.J. and Murray, B.A. (1992): Effects of antibodies against N-cadherin and N-CAM on the cranial neural crest and neural tube. Dev Biol, 153: 291–301
Bronner-Fraser, M. (2002): Molecular analysis of neural crest formation. J Physiol Paris, 96: 3–8
Brouns, M.R., Matheson, S.F., Hu, K.Q., Delalle, I., Caviness, V.S., Silver, J., Bronson, R.T. and Settleman, J. (2000): The adhesion signaling molecule p190 RhoGAP is required for morphogenetic processes in neural development. Development, 127: 4891–4903
Brown, S.A., Warburton, D., Brown, L.Y., Yu, C.Y., Roeder, E.R., Stengel-Rutkowski, S., Hennekam, R.C. and Muenke, M. (1998): Holoprosencephaly due to mutations in ZIC2, a homologue of Drosophila odd-paired. Nat Genet, 20: 180–183
Brown, L.Y., Odent, S., David, V., Blayau, M., Dubourg, C., Apacik, C., Delgado, M.A., Hall, B.D., Reynolds, J.F., Sommer, A., Wieczorek, D., Brown, S.A. and Muenke, M. (2001): Holoprosencephaly due to mutations in ZIC2: alanine tract expansion mutations may be caused by parental somatic recombination. Hum Mol Genet, 10: 791–796
Brunet, C.L., Sharpe, P.M. and Ferguson, M.W. (1995): Inhibition of TGF-beta 3 (but not TGF-beta 1 or TGF-beta 2) activity prevents normal mouse embryonic palate fusion. Int J Dev Biol, 39: 345–355
Buckingham, M. (2001): Skeletal muscle formation in vertebrates. Curr Opin Genet Dev, 11: 440–448
Burdi, A.R. and Silvey, R.G. (1969a): The relation of sex-associated facial profile reversal and stages of human palatal closure. Teratology, 2: 297–303
Burdi, A.R. and Silvey, R.G. (1969b): Sexual differences in closure of the human palatal shelves. Cleft Palate J, 6: 1–7

Cainarca, S., Messali, S., Ballabio, A. and Meroni, G. (1999): Functional characterization of the Opitz syndrome gene product (midin): evidence for homodimerization and association with microtubules throughout the cell cycle. Hum Mol Genet, 8: 1387–1396

Cairns, C. and McStay, B. (1995): Identification and cDNA cloning of a Xenopus nucleolar phosphoprotein, xNopp180, that is the homolog of the rat nucleolar protein Nopp140. J Cell Sci, 108: 3339–3347

Camus, A., Davidson, B.P., Billiards, S., Khoo, P., Rivera-Perez, J.A., Wakamiya, M., Behringer, R.R. and Tam, P.P. (2000): The morphogenetic role of midline mesendoderm and ectoderm in the development of the forebrain and the midbrain of the mouse embryo. Development, 127: 1799–1813

Cano, A., Perez-Moreno, M.A., Rodrigo, I., Locascio, A., Blanco, M.J., del Barrio, M.G., Portillo, F. and Nieto, M.A. (2000): The transcription factor snail controls epithelial-mesenchymal transitions by repressing E-cadherin expression. Nat Cell Biol, 2: 76–83

Capdevila, J., Tsukui, T., Rodriquez Esteban, C., Zappavigna, V. and Izpisua Belmonte, J.C. (1999): Control of vertebrate limb outgrowth by the proximal factor Meis2 and distal antagonism of BMPs by Gremlin. Mol Cell, 4: 839–849

Carette, M.J. and Ferguson, M.W. (1992): The fate of medial edge epithelial cells during palatal fusion in vitro: an analysis by DiI labelling and confocal microscopy. Development, 114: 379–388

Carl, T.F., Dufton, C., Hanken, J. and Klymkowsky, M.W. (1999): Inhibition of neural crest migration in Xenopus using antisense slug RNA. Dev Biol, 213: 101–115

Carter, M.G., Johns, M.A., Zeng, X., Zhou, L., Zink, M.C., Mankowski, J.L., Donovan, D.M. and Baylin, S.B. (2000): Mice deficient in the candidate tumor suppressor gene Hic1 exhibit developmental defects of structures affected in the Miller-Dieker syndrome. Hum Mol Genet, 9: 413–419

Caspary, T., Cleary, M.A., Perlman, E.J., Zhang, P., Elledge, S.J. and Tilghman, S.M. (1999): Oppositely imprinted genes p57(Kip2) and igf2 interact in a mouse model for Beckwith-Wiedemann syndrome. Genes Dev, 13: 3115–3124

Caton, A., Hacker, A., Naeem, A., Livet, J., Maina, F., Bladt, F., Klein, R., Birchmeier, C. and Guthrie, S. (2000): The branchial arches and HGF are growth-promoting and chemoattractant for cranial motor axons. Development, 127: 1751–1766

Carvajal, J.J., Cox, D., Summerbell, D. and Rigby, P.W. (2001): A BAC transgenic analysis of the Mrf4/Myf5 locus reveals interdigitated elements that control activation and maintenance of gene expression during muscle development. Development, 128: 1857–1868

Cecconi, F., Alvarez-Bolado, G., Meyer, B.I., Roth, K.A. and Gruss, P. (1998): Apaf1 (CED-4 homolog) regulates programmed cell death in mammalian development. Cell, 94: 727–737

Celli, G., LaRochelle, W.J., Mackem, S., Sharp, R. and Merlino, G. (1998): Soluble dominant-negative receptor uncovers essential roles for fibroblast growth factors in multi-organ induction and patterning. Embo J, 17: 1642–1655

Chai, Y., Sasano, Y., Bringas, P., Jr., Mayo, M., Kaartinen, V., Heisterkamp, N., Groffen, J., Slavkin, H. and Shuler, C. (1997): Characterization of the fate of midline epithelial cells during the fusion of mandibular prominences in vivo. Dev Dyn, 208: 526–535

Chai, Y., Jiang, X., Ito, Y., Bringas, P., Jr., Han, J., Rowitch, D.H., Soriano, P., McMahon, A.P. and Sucov, H.M. (2000): Fate of the mammalian cranial neural crest during tooth and mandibular morphogenesis. Development, 127: 1671–1679

Chalepakis, G., Stoykova, A., Wijnholds, J., Tremblay, P. and Gruss, P. (1993): Pax: gene regulators in the developing nervous system. J Neurobiol, 24: 1367–1384

Chalepakis, G., Wijnholds, J., Giese, P., Schachner, M. and Gruss, P. (1994): Characterization of Pax-6 and Hoxa-1 binding to the promoter region of the neural cell adhesion molecule L1. DNA Cell Biol, 13: 891–900

Chambers, D. and Mason, I. (2000): Expression of sprouty2 during early development of the chick embryo is coincident with known sites of FGF signalling. Mech Dev, 91: 361–364

Chan, W.Y. and Tam, P.P. (1988): A morphological and experimental study of the mesencephalic neural crest cells in the mouse embryo using wheat germ agglutinin-gold conjugate as the cell marker. Development, 102: 427–442

Chan, T.S., Lin, C.X., Chan, W.Y., Chung, S.S. and Chung, S.K. (1995): Mouse preproendothelin-1 gene. cDNA cloning, sequence analysis and determination of sites of expression during embryonic development. Eur J Biochem, 234: 819–826

Chan, J., Mably, J.D., Serluca, F.C., Chen, J.N., Goldstein, N.B., Thomas, M.C., Cleary, J.A., Brennan, C., Fishman, M.C. and Roberts, T.M. (2001): Morphogenesis of prechordal plate and notochord requires intact Eph/ephrin B signaling. Dev Biol, 234: 470–482
Chapman, D.L., Garvey, N., Hancock, S., Alexiou, M., Agulnik, S.I., Gibson-Brown, J.J., Cebra-Thomas, J., Bollag, R.J., Silver, L.M. and Papaioannou, V.E. (1996): Expression of the T-box family genes, Tbx1-Tbx5, during early mouse development. Dev Dyn, 206: 379–390
Charité, J., McFadden, D.G., Merlo, G., Levi, G., Clouthier, D.E., Yanagisawa, M., Richardson, J.A. and Olson, E.N. (2001): Role of Dlx6 in regulation of an endothelin-1-dependent, dHAND branchial arch enhancer. Genes Dev, 15: 3039–3049
Chen, Z.F. and Behringer, R.R. (1995): twist is required in head mesenchyme for cranial neural tube morphogenesis. Genes Dev, 9: 686–699
Chen, J., Chang, S., Duncan, S.A., Okano, H.J., Fishell, G. and Aderem, A. (1996): Disruption of the MacMARCKS gene prevents cranial neural tube closure and results in anencephaly. Proc Natl Acad Sci U S A, 93: 6275–6279
Chiang, C., Litingtung, Y., Lee, E., Young, K.E., Corden, J.L., Westphal, H. and Beachy, P.A. (1996): Cyclopia and defective axial patterning in mice lacking Sonic hedgehog gene function. Nature, 383: 407–413
Chin, J.R. and Werb, Z. (1997): Matrix metalloproteinases regulate morphogenesis, migration and remodeling of epithelium, tongue skeletal muscle and cartilage in the mandibular arch. Development, 124: 1519–1530
Choi, D.S., Ward, S.J., Messaddeq, N., Launay, J.M. and Maroteaux, L. (1997): 5-HT2B receptor-mediated serotonin morphogenetic functions in mouse cranial neural crest and myocardiac cells. Development, 124: 1745–1755
Clifton-Bligh, R.J., Wentworth, J.M., Heinz, P., Crisp, M.S., John, R., Lazarus, J.H., Ludgate, M. and Chatterjee, V.K. (1998): Mutation of the gene encoding human TTF-2 associated with thyroid agenesis, cleft palate and choanal atresia. Nat Genet, 19: 399–401
Clouthier, D.E., Hosoda, K., Richardson, J.A., Williams, S.C., Yanagisawa, H., Kuwaki, T., Kumada, M., Hammer, R.E. and Yanagisawa, M. (1998): Cranial and cardiac neural crest defects in endothelin-A receptor- deficient mice. Development, 125: 813–824
Clouthier, D.E., Williams, S.C., Yanagisawa, H., Wieduwilt, M., Richardson, J.A. and Yanagisawa, M. (2000): Signaling pathways crucial for craniofacial development revealed by endothelin-A receptor-deficient mice. Dev Biol, 217: 10–24
Coffin, J.D. and Poole, T.J. (1988): Embryonic vascular development: immunohistochemical identification of the origin and subsequent morphogenesis of the major vessel primordia in quail embryos. Development, 102: 735–748
Cohen, M.M., Jr. (1989a): Perspectives on holoprosencephaly: Part I. Epidemiology, genetics, and syndromology. Teratology, 40: 211–235
Cohen, M.M., Jr. (1989b): Perspectives on holoprosencephaly: Part III. Spectra, distinctions, continuities, and discontinuities. Am J Med Genet, 34: 271–288
Cohen, M.M., Jr. and Sulik, K.K. (1992): Perspectives on holoprosencephaly: Part II. Central nervous system, craniofacial anatomy, syndrome commentary, diagnostic approach, and experimental studies. J Craniofac Genet Dev Biol, 12: 196–244
Condie, B.G., Bain, G., Gottlieb, D.I. and Capecchi, M.R. (1997): Cleft palate in mice with a targeted mutation in the gamma-aminobutyric acid-producing enzyme glutamic acid decarboxylase 67. Proc Natl Acad Sci U S A, 94: 11451–11455
Conlon, F.L., Lyons, K.M., Takaesu, N., Barth, K.S., Kispert, A., Herrmann, B. and Robertson, E.J. (1994): A primary requirement for nodal in the formation and maintenance of the primitive streak in the mouse. Development, 120: 1919–1928
Conway, S.J., Henderson, D.J. and Copp, A.J. (1997a): Pax3 is required for cardiac neural crest migration in the mouse: evidence from the splotch (Sp2H) mutant. Development, 124: 505–514
Conway, S.J., Henderson, D.J., Kirby, M.L., Anderson, R.H. and Copp, A.J. (1997b): Development of a lethal congenital heart defect in the splotch (Pax3) mutant mouse. Cardiovasc Res, 36: 163–173
Conway, S.J., Bundy, J., Chen, J., Dickman, E., Rogers, R. and Will, B.M. (2000): Decreased neural crest stem cell expansion is responsible for the conotruncal heart defects within the splotch (Sp(2H))/Pax3 mouse mutant. Cardiovasc Res, 47: 314–328

Corbel, C., Lehmann, A., Davison, F. (2000): Expression of CD44 during early development of the chick embryo. Mech Dev, 96:111–4
Couly, G.F. and Le Douarin, N.M. (1985): Mapping of the early neural primordium in quail-chick chimeras. I. Developmental relationships between placodes, facial ectoderm, and prosencephalon. Dev Biol, 110: 422–439
Couly, G.F. and Le Douarin, N.M. (1987): Mapping of the early neural primordium in quail-chick chimeras. II. The prosencephalic neural plate and neural folds: implications for the genesis of cephalic human congenital abnormalities. Dev Biol, 120: 198–214
Couly, G. and Le Douarin, N.M. (1990): Head morphogenesis in embryonic avian chimeras: evidence for a segmental pattern in the ectoderm corresponding to the neuromeres. Development, 108: 543–558
Couly, G.F., Coltey, P.M. and Le Douarin, N.M. (1992): The developmental fate of the cephalic mesoderm in quail-chick chimeras. Development, 114: 1–15
Couly, G.F., Coltey, P.M. and Le Douarin, N.M. (1993): The triple origin of skull in higher vertebrates: a study in quail- chick chimeras. Development, 117: 409–429
Couly, G., Coltey, P., Eichmann, A. and Le Douarin, N.M. (1995): The angiogenic potentials of the cephalic mesoderm and the origin of brain and head blood vessels. Mech Dev, 53: 97–112
Couly, G., Grapin-Botton, A., Coltey, P. and Le Douarin, N.M. (1996): The regeneration of the cephalic neural crest, a problem revisited: the regenerating cells originate from the contralateral or from the anterior and posterior neural fold. Development, 122: 3393-3407
Couly, G., Grapin-Botton, A., Coltey, P., Ruhin, B. and Le Douarin, N.M. (1998): Determination of the identity of the derivatives of the cephalic neural crest: incompatibility between Hox gene expression and lower jaw development. Development, 125: 3445–3459
Couly, G., Creuzet, S., Bennaceur, S., Vincent, C. and Le Douarin, N.M. (2002): Interactions between Hox-negative cephalic neural crest cells and the foregut endoderm in patterning the facial skeleton in the vertebrate head. Development, 129: 1061–1073
Coventry, S., Kapur, R.P. and Siebert, J.R. (1998): Cyclopamine-induced holoprosencephaly and associated craniofacial malformations in the golden hamster: anatomic and molecular events. Pediatr Dev Pathol, 1: 29–41
Cox, G.A., Mahaffey, C.L., Nystuen, A., Letts, V.A. and Frankel, W.N. (2000): The mouse fidgetin gene defines a new role for AAA family proteins in mammalian development. Nat Genet, 26: 198–202.
Crosnier, C., Driancourt, C., Raynaud, N., Hadchouel, M. and Meunier-Rotival, M. (2001): Fifteen novel mutations in the JAGGED1 gene of patients with Alagille syndrome. Hum Mutat, 17: 72–73
Cuervo, R., Valencia, C., Chandraratna, R.A.S. and Covarrubias, L. (2002): Programmed cell death is required for palate shelf fusion and is regulated by retinoic acid. Dev Biol, 245: 145–156.
Dal Zotto, L., Quaderi, N.A., Elliott, R., Lingerfelter, P.A., Carrel, L., Valsecchi, V., Montini, E., Yen, C.H., Chapman, V., Kalcheva, I., Arrigo, G., Zuffardi, O., Thomas, S., Willard, H.F., Ballabio, A., Disteche, C.M. and Rugarli, E.I. (1998): The mouse Mid1 gene: implications for the pathogenesis of Opitz syndrome and the evolution of the mammalian pseudoautosomal region. Hum Mol Genet, 7: 489–499
D'Amico-Martel, A. and Noden, D.M. (1983): Contributions of placodal and neural crest cells to avian cranial peripheral ganglia. Am J Anat, 166: 445–468
Dale, J.K., Vesque, C., Lints, T.J., Sampath, T.K., Furley, A., Dodd, J. and Placzek, M. (1997): Cooperation of BMP7 and SHH in the induction of forebrain ventral midline cells by prechordal mesoderm. Cell, 90: 257–269
Datson, N.A., Semina, E., van Staalduinen, A.A., Dauwerse, H.G., Meershoek, E.J., Heus, J.J., Frants, R.R., den Dunnen, J.T., Murray, J.C. and van Ommen, G.J. (1996): Closing in on the Rieger syndrome gene on 4q25: mapping translocation breakpoints within a 50-kb region. Am J Hum Genet, 59: 1297–1305
Dattani, M.T., Martinez-Barbera, J.P., Thomas, P.Q., Brickman, J.M., Gupta, R., Martensson, I.L., Toresson, H., Fox, M., Wales, J.K., Hindmarsh, P.C., Krauss, S., Beddington, R.S. and Robinson, I.C. (1998): Mutations in the homeobox gene HESX1/Hesx1 associated with septo-optic dysplasia in human and mouse. Nat Genet, 19: 125–133
Daw, S.C., Taylor, C., Kraman, M., Call, K., Mao, J., Schuffenhauer, S., Meitinger, T., Lipson, T., Goodship, J. and Scambler, P. (1996): A common region of 10p deleted in DiGeorge and velocardiofacial syndromes. Nat Genet, 13: 458–460

Deardorff, M.A., Tan, C., Saint-Jeannet, J.P. and Klein, P.S. (2001): A role for frizzled 3 in neural crest development. Development, 128: 3655–3663

De Felice, M., Ovitt, C., Biffali, E., Rodriguez-Mallon, A., Arra, C., Anastassiadis, K., Macchia, P.E., Mattei, M.G., Mariano, A., Scholer, H., Macchia, V. and Di Lauro, R. (1998): A mouse model for hereditary thyroid dysgenesis and cleft palate. Nat Genet, 19: 395–398

Degenhardt, K. and Sassoon, D.A. (2001): A role for Engrailed-2 in determination of skeletal muscle physiologic properties. Dev Biol, 231: 175–189

Degenhardt, K., Rentschler, S., Fishman, G. and Sassoon, D.A. (2002): Cellular and cis-regulation of *En-2* expression in the mandibular arch. Mech Dev, 111: 125–136

De Maximy, A.A., Nakatake, Y., Moncada, S., Itoh, N., Thiery, J.P. and Bellusci, S. (1999): Cloning and expression pattern of a mouse homologue of drosophila sprouty in the mouse embryo. Mech Dev, 81: 213–216

De Moerlooze, L., Spencer-Dene, B., Revest, J., Hajihosseini, M., Rosewell, I. and Dickson, C. (2000): An important role for the IIIb isoform of fibroblast growth factor receptor 2 (FGFR2) in mesenchymal-epithelial signalling during mouse organogenesis. Development, 127: 483–492

De Wals, P., Bloch, D., Calabro, A., Calzolari, E., Cornel, M.C., Johnson, Z., Ligutic, I., Nevin, N., Pexieder, I., Stoll, C. and et al. (1991): Association between holoprosencephaly and exposure to topical retinoids: results of the EUROCAT Survey. Paediatr Perinat Epidemiol, 5: 445–447

Depew, M.J., Liu, J.K., Long, J.E., Presley, R., Meneses, J.J., Pedersen, R.A. and Rubenstein, J.L. (1999): Dlx5 regulates regional development of the branchial arches and sensory capsules. Development, 126: 3831–3846

Derry, J.M., Gormally, E., Means, G.D., Zhao, W., Meindl, A., Kelley, R.I., Boyd, Y. and Herman, G.E. (1999): Mutations in a delta 8-delta 7 sterol isomerase in the tattered mouse and X-linked dominant chondrodysplasia punctata. Nat Genet, 22: 286–290

Dixon, J.E. and Kintner, C.R. (1989): Cellular contacts required for neural induction in Xenopus embryos: evidence for two signals. Development, 106: 749–757

Dixon, J., Edwards, S.J., Anderson, I., Brass, A., Scambler, P.J. and Dixon, M.J. (1997a): Identification of the complete coding sequence and genomic organization of the Treacher Collins syndrome gene. Genome Res, 7: 223–234

Dixon, J., Hovanes, K., Shiang, R. and Dixon, M.J. (1997b): Sequence analysis, identification of evolutionary conserved motifs and expression analysis of murine tcof1 provide further evidence for a potential function for the gene and its human homologue, TCOF1. Hum Mol Genet, 6: 727–737

Dixon, M. and Lumsden, A. (1999): Distribution of neuregulin-1 (nrg1) and erbB4 transcripts in embryonic chick hindbrain. Mol Cell Neurosci, 13: 237–258

Dixon, J., Brakebusch, C., Fassler, R. and Dixon, M.J. (2000): Increased levels of apoptosis in the prefusion neural folds underlie the craniofacial disorder, Treacher Collins syndrome. Hum Mol Genet, 9: 1473–1480

Donohoe, M.E., Zhang, X., McGinnis, L., Biggers, J., Li, E. and Shi, Y. (1999): Targeted disruption of mouse Yin Yang 1 transcription factor results in peri-implantation lethality. Mol Cell Biol, 19: 7237–7244

Duband, J.L. and Thiery, J.P. (1987): Distribution of laminin and collagens during avian neural crest development. Development, 101: 461–478

Duband, J.L., Monier, F., Delannet, M. and Newgreen, D. (1995): Epithelium-mesenchyme transition during neural crest development. Acta Anat, 154: 63–78

Dudley, A.T., Lyons, K.M. and Robertson, E.J. (1995): A requirement for bone morphogenetic protein-7 during development of the mammalian kidney and eye. Genes Dev, 9: 2795–2807

Duncan, M.K., Kozmik, Z., Cveklova, K., Piatigorsky, J. and Cvekl, A. (2000): Overexpression of PAX6(5a) in lens fiber cells results in cataract and upregulation of (alpha)5(beta)1 integrin expression. J Cell Sci, 113: 3173–3185

Dunn, N.R., Winnier, G.E., Hargett, L.K., Schrick, J.J., Fogo, A.B. and Hogan, B.L. (1997): Haploinsufficient phenotypes in Bmp4 heterozygous null mice and modification by mutations in Gli3 and Alx4. Dev Biol, 188: 235–247

Dupe, V., Ghyselinck, N.B., Thomazy, V., Nagy, L., Davies, P.J., Chambon, P. and Mark, M. (1999): Essential roles of retinoic acid signaling in interdigital apoptosis and control of BMP-7 expression in mouse autopods. Dev Biol, 208: 30–43

Edelman, G.M. and Jones, F.S. (1998): Gene regulation of cell adhesion: a key step in neural morphogenesis. Brain Res Brain Res Rev, 26: 337–352.

Edery, P., Attie, T., Amiel, J., Pelet, A., Eng, C., Hofstra, R.M., Martelli, H., Bidaud, C., Munnich, A. and Lyonnet, S. (1996): Mutation of the endothelin-3 gene in the Waardenburg-Hirschsprung disease (Shah-Waardenburg syndrome). Nat Genet, 12: 442–444

Edwards, S.J., Gladwin, A.J. and Dixon, M.J. (1997): The mutational spectrum in Treacher Collins syndrome reveals a predominance of mutations that create a premature-termination codon. Am J Hum Genet, 60: 515–524

Eichmann, A., Marcelle, C., Breant, C. and Le Douarin, N.M. (1993): Two molecules related to the VEGF receptor are expressed in early endothelial cells during avian embryonic development. Mech Dev, 42: 33–48

Eichmann, A., Marcelle, C., Breant, C. and Le Douarin, N.M. (1996): Molecular cloning of Quek 1 and 2, two quail vascular endothelial growth factor (VEGF) receptor-like molecules. Gene, 174: 3–8

Eickholt, B.J., Mackenzie, S.L., Graham, A., Walsh, F.S. and Doherty, P. (1999): Evidence for collapsin-1 functioning in the control of neural crest migration in both trunk and hindbrain regions. Development, 126: 2181–2189

Ekker, S.C., Ungar, A.R., Greenstein, P., von Kessler, D.P., Porter, J.A., Moon, R.T. and Beachy, P.A. (1995): Patterning activities of vertebrate hedgehog proteins in the developing eye and brain. Curr Biol, 5: 944–955

El Ghouzzi, V., Le Merrer, M., Perrin-Schmitt, F., Lajeunie, E., Benit, P., Renier, D., Bourgeois, P., Bolcato-Bellemin, A.L., Munnich, A. and Bonaventure, J. (1997): Mutations of the TWIST gene in the Saethre-Chotzen syndrome. Nat Genet, 15: 42–46

El Ghouzzi, V., Lajeunie, E., Le Merrer, M., Cormier-Daire, V., Renier, D., Munnich, A. and Bonaventure, J. (1999): Mutations within or upstream of the basic helix-loop-helix domain of the TWIST gene are specific to Saethre-Chotzen syndrome. Eur J Hum Genet, 7: 27–33

El Ghouzzi, V., Legeai-Mallet, L., Benoist-Lasselin, C., Lajeunie, E., Renier, D., Munnich, A. and Bonaventure, J. (2001): Mutations in the basic domain and the loop-helix II junction of TWIST abolish DNA binding in Saethre-Chotzen syndrome. FEBS Lett, 492: 112–118

Ellies, D.L., Church, V., Francis-West, P. and Lumsden, A. (2000): The WNT antagonist cSFRP2 modulates programmed cell death in the developing hindbrain. Development, 127: 5285–5295

Endo, Y., Osumi, N. and Wakamatsu, Y. (2002): Bimodal functions of notch-mediated signaling are involved in neural crest formation during avian ectoderm development. Development, 129: 863–873

Epperlein, H._H. and Lehman, R. (1975): The ectomesenchymal-endodermal interaction system (EEIS) of *Triturus alpestris* in tissue culture. 2. Obervations on differentiation of visceral cartilage. Differentiation, 4: 159–174.

Epstein, D.J., Vogan, K.J., Trasler, D.G. and Gros, P. (1993): A mutation within intron 3 of the Pax-3 gene produces aberrantly spliced mRNA transcripts in the splotch (Sp) mouse mutant. Proc Natl Acad Sci U S A, 90: 532–536

Epstein, J.A., Li, J., Lang, D., Chen, F., Brown, C.B., Jin, F., Lu, M.M., Thomas, M., Liu, E., Wessels, A. and Lo, C.W. (2000): Migration of cardiac neural crest cells in Splotch embryos. Development, 127: 1869–1878

Erickson, C.A. and Perris, R. (1993): The role of cell-cell and cell-matrix interactions in the morphogenesis of the neural crest. Dev Biol, 159: 60–74

Etchevers, H.C., Couly, G., Vincent, C. and Le Douarin, N.M. (1999): Anterior cephalic neural crest is required for forebrain viability. Development, 126: 3533–3543

Etchevers, H.C., Vincent, C., Le Douarin, N.M. and Couly, G.F. (2001): The cephalic neural crest provides pericytes and smooth muscle cells to all blood vessels of the face and forebrain. Development, 128: 1059–1068

Ewart, J.L., Cohen, M.F., Meyer, R.A., Huang, G.Y., Wessels, A., Gourdie, R.G., Chin, A.J., Park, S.M., Lazatin, B.O., Villabon, S. and Lo, C.W. (1997): Heart and neural tube defects in transgenic mice overexpressing the Cx43 gap junction gene. Development, 124: 1281–1292

Fang, J. and Hall, B.K. (1997): Chondrogenic cell differentiation from membrane bone periostea. Anat Embryol (Berl), 196: 349–362

Farlie, P.G., Kerr, R., Thomas, P., Symes, T., Minichiello, J., Hearn, C.J. and Newgreen, D. (1999): A paraxial exclusion zone creates patterned cranial neural crest cell outgrowth adjacent to rhombomeres 3 and 5. Dev Biol, 213: 70–84

Farrell, M.J., Stadt, H., Wallis, K.T., Scambler, P., Hixon, R.L., Wolfe, R., Leatherbury, L. and Kirby, M.L. (1999): HIRA, a DiGeorge syndrome candidate gene, is required for cardiac outflow tract septation. Circ Res, 84: 127–135

Feldman, B., Gates, M.A., Egan, E.S., Dougan, S.T., Rennebeck, G., Sirotkin, H.I., Schier, A.F. and Talbot, W.S. (1998): Zebrafish organizer development and germ-layer formation require nodal-related signals. Nature, 395: 181–185

Feldman, B., Dougan, S.T., Schier, A.F. and Talbot, W.S. (2000): Nodal-related signals establish mesendodermal fate and trunk neural identity in zebrafish. Curr Biol, 10: 531–534.

Ferguson, M.W. (1988): Palate development. Development, 103: 41–60

Ferguson, C.A., Tucker, A.S. and Sharpe, P.T. (2000): Temporospatial cell interactions regulating mandibular and maxillary arch patterning. Development, 127: 403–412

Fitchett, J.E. and Hay, E.D. (1989): Medial edge epithelium transforms to mesenchyme after embryonic palatal shelves fuse. Dev Biol, 131: 455–474

Fitzpatrick, D.R., Denhez, F., Kondaiah, P. and Akhurst, R.J. (1990): Differential expression of TGF beta isoforms in murine palatogenesis. Development, 109: 585–595

Fleming, A. and Copp, A.J. (2000): A genetic risk factor for mouse neural tube defects: defining the embryonic basis. Hum Mol Genet, 9: 575–581

Flenniken, A.M., Gale, N.W., Yancopoulos, G.D. and Wilkinson, D.G. (1996): Distinct and overlapping expression patterns of ligands for Eph-related receptor tyrosine kinases during mouse embryogenesis. Dev Biol, 179: 382–401

Flomen, R.H., Vatcheva, R., Gorman, P.A., Baptista, P.R., Groet, J., Barisic, I., Ligutic, I. and Nizetic, D. (1998): Construction and analysis of a sequence-ready map in 4q25: Rieger syndrome can be caused by haploinsufficiency of RIEG, but also by chromosome breaks approximately 90kb upstream of this gene. Genomics, 47: 409–413

Foley, A.C., Storey, K.G. and Stern, C.D. (1997): The prechordal region lacks neural inducing ability, but can confer anterior character to more posterior neuroepithelium. Development, 124: 2983–2996

Francis-West, P.H., Tatla, T. and Brickell, P.M. (1994): Expression patterns of the bone morphogenetic protein genes Bmp-4 and Bmp-2 in the developing chick face suggest a role in outgrowth of the primordia. Dev Dyn, 201: 168–178

Francis-West, P., Ladher, R., Barlow, A. and Graveson, A. (1998): Signalling interactions during facial development. Mech Dev, 75: 3–28

Francis-West, P.H., Ladher, R. and Schoenwolf, G. (2002): Development of the sensory organs. Frontiers in Science. 85: 151–173

Fraser, S., Keynes, R. and Lumsden, A. (1990): Segmentation in the chick embryo hindbrain is defined by cell lineage restrictions. Nature, 344: 431–435

Frenz, D.A. and Van De Water, T.R. (1991): Epithelial control of periotic mesenchyme chondrogenesis. Dev Biol, 144: 38–46

Frenz, D.A., Galinovic-Schwartz, V., Liu, W., Flanders, K.C. and Van de Water, T.R. (1992): Transforming growth factor beta 1 is an epithelial-derived signal peptide that influences otic capsule formation. Dev Biol, 153: 324–336

Frenz, D.A., Liu, W., Williams, J.D., Hatcher, V., Galinovic-Schwartz, V., Flanders, K.C. and Van de Water, T.R. (1994): Induction of chondrogenesis: requirement for synergistic interaction of basic fibroblast growth factor and transforming growth factor-beta. Development, 120: 415–424

Frisch, S.M. and Francis, H. (1994): Disruption of epithelial cell-matrix interactions induces apoptosis. J Cell Biol, 124: 619–626

Füchtbauer, E.M. (1995): Expression of M-twist during postimplantation development of the mouse. Dev Dyn, 204: 316–322

Fujiwara, M., Uchida, T., Osumi-Yamashita, N. and Eto, K. (1994): Uchida rat (rSey): a new mutant rat with craniofacial abnormalities resembling those of the mouse Sey mutant. Differentiation, 57: 31–38

Fukiishi, Y. and Morriss-Kay, G.M. (1992): Migration of cranial neural crest cells to the pharyngeal arches and heart in rat embryos. Cell Tissue Res, 268: 1–8

Funahashi, J., Sekido, R., Murai, K., Kamachi, Y. and Kondoh, H. (1993): Delta-crystallin enhancer binding protein delta EF1 is a zinc finger- homeodomain protein implicated in postgastrulation embryogenesis. Development, 119: 433–446

Furuta, Y., Piston, D.W. and Hogan, B.L. (1997): Bone morphogenetic proteins (BMPs) as regulators of dorsal forebrain development. Development, 124: 2203–2212

Fyfe, D.M. and Hall, B.K. (1979): Lack of association between avian cartilages of different embryological origins when maintained in vitro. Am J Anat, 154: 485–496

Gage, P.J. and Camper, S.A. (1997): Pituitary homeobox 2, a novel member of the bicoid-related family of homeobox genes, is a potential regulator of anterior structure formation. Hum Mol Genet, 6: 457–464

Gage, P.J., Suh, H. and Camper, S.A. (1999): Dosage requirement of Pitx2 for development of multiple organs. Development, 126: 4643–4651

Garg, V., Yamagishi, C., Hu, T., Kathiriya, I.S., Yamagishi, H. and Srivastava, D. (2001): Tbx1, a DiGeorge syndrome candidate gene, is regulated by sonic hedgehog during pharyngeal arch development. Dev Biol, 235: 62–73

Gassmann, M., Casagranda, F., Orioli, D., Simon, H., Lai, C., Klein, R. and Lemke, G. (1995): Aberrant neural and cardiac development in mice lacking the ErbB4 neuregulin receptor. Nature, 378: 390–394

Gavalas, A. and Krumlauf, R. (2000): Retinoid signalling and hindbrain patterning. Curr Opin Genet Dev, 10: 380–386

Gavalas, A., Trainor, P., Ariza-McNaughton, L. and Krumlauf, R. (2001): Synergy between Hoxa1 and Hoxb1: the relationship between arch patterning and the generation of cranial neural crest. Development, 128: 3017–3027

Geelen, J.A. and Langman, J. (1977): Closure of the neural tube in the cephalic region of the mouse embryo. Anat Rec, 189: 625–640

Gendron-Maguire, M., Mallo, M., Zhang, M. and Gridley, T. (1993): Hoxa-2 mutant mice exhibit homeotic transformation of skeletal elements derived from cranial neural crest. Cell, 75: 1317–1331

Gerritsen, M.E., Williams, A.J., Neish, A.S., Moore, S., Shi, Y. and Collins, T. (1997): CREB-binding protein/p300 are transcriptional coactivators of p65. Proc Natl Acad Sci U S A, 94: 2927–2932

Ghyselinck, N.B., Dupe, V., Dierich, A., Messaddeq, N., Garnier, J.M., Rochette-Egly, C., Chambon, P. and Mark, M. (1997): Role of the retinoic acid receptor beta (RARbeta) during mouse development. Int J Dev Biol, 41: 425–447

Gitelman, I. (1997): Twist protein in mouse embryogenesis. Dev Biol, 189: 205–214

Gladwin, A.J., Dixon, J., Loftus, S.K., Edwards, S., Wasmuth, J.J., Hennekam, R.C. and Dixon, M.J. (1996): Treacher Collins syndrome may result from insertions, deletions or splicing mutations, which introduce a termination codon into the gene. Hum Mol Genet, 5: 1533–1538

Glinka, A., Wu, W., Onichtchouk, D., Blumenstock, C. and Niehrs, C. (1997): Head induction by simultaneous repression of Bmp and Wnt signalling in Xenopus. Nature, 389: 517–519

Glinka, A., Wu, W., Delius, H., Monaghan, A.P., Blumenstock, C. and Niehrs, C. (1998): Dickkopf-1 is a member of a new family of secreted proteins and functions in head induction. Nature, 391: 357–362

Goh, K.L., Yang, J.T. and Hynes, R.O. (1997): Mesodermal defects and cranial neural crest apoptosis in alpha5 integrin-null embryos. Development, 124: 4309–4319

Golden, J.A. and Chernoff, G.F. (1995): Multiple sites of anterior neural tube closure in humans: evidence from anterior neural tube defects (anencephaly). Pediatrics, 95: 506–510

Golden, J.A., Bracilovic, A., McFadden, K.A., Beesley, J.S., JL, R.R. and Grinspan, J.B. (1999): Ectopic bone morphogenetic proteins 5 and 4 in the chicken forebrain lead to cyclopia and holoprosencephaly. Proc Natl Acad Sci U S A, 96: 2439–2444

Golding, J.P., Tidcombe, H., Tsoni, S. and Gassmann, M. (1999): Chondroitin sulphate-binding molecules may pattern central projections of sensory axons within the cranial mesenchyme of the developing mouse. Dev Biol, 216: 85–97

Golding, J.P., Trainor, P., Krumlauf, R. and Gassmann, M. (2000): Defects in pathfinding by cranial neural crest cells in mice lacking the neuregulin receptor ErbB4. Nat Cell Biol, 2: 103–109

Goldmuntz, E. and Emanuel, B.S. (1997): Genetic disorders of cardiac morphogenesis. The DiGeorge and velocardiofacial syndromes. Circ Res, 80: 437–443

Gong, W., Emanuel, B.S., Collins, J., Kim, D.H., Wang, Z., Chen, F., Zhang, G., Roe, B. and Budarf, M.L. (1996): A transcription map of the DiGeorge and velo-cardio-facial syndrome minimal critical region on 22q11. Hum Mol Genet, 5: 789–800

Gomez-Skarmeta, J.L., de la Calle-Mustienes, E., Modolell, J. and Mayor, R. (1999): Xenopus brain factor-2 controls mesoderm, forebrain and neural crest development. Mech Dev, 80: 15–27

Gorski, J.L., Estrada, L., Hu, C. and Liu, Z. (2000): Skeletal-specific expression of Fgd1 during bone formation and skeletal defects in faciogenital dysplasia (FGDY; Aarskog syndrome). Dev Dyn, 218: 573–586

Gould, A., Itasaki, N. and Krumlauf, R. (1998): Initiation of rhombomeric Hoxb4 expression requires induction by somites and a retinoid pathway. Neuron, 21: 39–51

Graham, A., Papalopulu, N. and Krumlauf, R. (1989): The murine and Drosophila homeobox gene complexes have common features of organization and expression. Cell, 57: 367–378

Graham, A., Heyman, I. and Lumsden, A. (1993): Even-numbered rhombomeres control the apoptotic elimination of neural crest cells from odd-numbered rhombomeres in the chick hindbrain. Development, 119: 233–245

Graham, A., Francis-West, P., Brickell, P. and Lumsden, A. (1994): The signalling molecule BMP4 mediates apoptosis in the rhombencephalic neural crest. Nature, 372: 684–686

Graham, A., Koentges, G. and Lumsden, A. (1996): Neural crest apoptosis and the establishment of craniofacial pattern: an honorable death. Mol Cell Neurosci, 8: 76–83

Grammatopoulos, G.A., Bell, E., Toole, L., Lumsden, A. and Tucker, A.S. (2000): Homeotic transformation of branchial arch identity after Hoxa2 overexpression. Development, 127: 5355–5365

Grapin-Botton, A., Bonnin, M.A., McNaughton, L.A., Krumlauf, R. and Le Douarin, N.M. (1995): Plasticity of transposed rhombomeres: Hox gene induction is correlated with phenotypic modifications. Development, 121: 2707–2721

Grapin-Botton, A., Bonnin, M.A. and Le Douarin, N.M. (1997): Hox gene induction in the neural tube depends on three parameters: competence, signal supply and paralogue group. Development, 124: 849–859

Graveson, A.C. and Armstrong, J.B. (1987): Differentiation of cartilage from cranial neural crest in the axolotl (Ambystoma mexicanum). Differentiation, 35: 16–20

Green, J.B., New, H.V. and Smith, J.C. (1992): Responses of embryonic Xenopus cells to activin and FGF are separated by multiple dose thresholds and correspond to distinct axes of the mesoderm. Cell, 71: 731–739

Greene, R.M. and Pratt, R.M. (1977): Inhibition by diazo-oxo-norleucine (DON) of rat palatal glycoproteins synthesis and epithelial cell adhesion in vitro. Exp Cell Res, 105: 27–37

Greene, R.M., Shah, R.M., Lloyd, M.R., Crawford, B.J., Suen, R., Shanfeld, J.L. and Davidovitch, Z. (1983): Differentiation of the avian secondary palate. J Exp Zool, 225: 43–52

Greene, N.D., Gerrelli, D., Van Straaten, H.W. and Copp, A.J. (1998): Abnormalities of floor plate, notochord and somite differentiation in the loop-tail (Lp) mouse: a model of severe neural tube defects. Mech Dev, 73: 59–72

Griffith, C.M. and Hay, E.D. (1992): Epithelial-mesenchymal transformation during palatal fusion: carboxyfluorescein traces cells at light and electron microscopic levels. Development, 116: 1087–1099

Grindley, J.C., Davidson, D.R. and Hill, R.E. (1995): The role of Pax-6 in eye and nasal development. Development, 121: 1433–1442

Grindley, J.C., Hargett, L.K., Hill, R.E., Ross, A. and Hogan, B.L. (1997): Disruption of PAX6 function in mice homozygous for the Pax6Sey-1Neu mutation produces abnormalities in the early development and regionalization of the diencephalon. Mech Dev, 64: 111–126

Gripp, K.W., Wotton, D., Edwards, M.C., Roessler, E., Ades, L., Meinecke, P., Richieri-Costa, A., Zackai, E.H., Massague, J., Muenke, M. and Elledge, S.J. (2000): Mutations in TGIF cause holoprosencephaly and link NODAL signalling to human neural axis determination. Nat Genet, 25: 205–208.

Gritsman, K., Zhang, J., Cheng, S., Heckscher, E., Talbot, W.S. and Schier, A.F. (1999): The EGF-CFC protein one-eyed pinhead is essential for nodal signaling. Cell, 97: 121–132

Gritsman, K., Talbot, W.S. and Schier, A.F. (2000): Nodal signaling patterns the organizer. Development, 127: 921–932

Günther, T., Struwe, M., Aguzzi, A. and Schughart, K. (1994): Open brain, a new mouse mutant with severe neural tube defects, shows altered gene expression patterns in the developing spinal cord. Development, 120: 3119–3130

Guris, D.L., Fantes, J., Tara, D., Druker, B.J. and Imamoto, A. (2001): Mice lacking the homologue of the human 22q11.2 gene CRKL phenocopy neurocristopathies of DiGeorge syndrome. Nat Genet, 27: 293–298

Hacker, A. and Guthrie, S. (1998): A distinct developmental programme for the cranial paraxial mesoderm in the chick embryo. Development, 125: 3461–3472

Hackett, D.A., Smith, J.L. and Schoenwolf, G.C. (1997): Epidermal ectoderm is required for full elevation and for convergence during bending of the avian neural plate. Dev Dyn, 210: 397–406

Hadchouel, J., Tajbakhsh, S., Primig, M., Chang, T.H., Daubas, P., Rocancourt, D. and Buckingham, M. (2000): Modular long-range regulation of Myf5 reveals unexpected heterogeneity between skeletal muscles in the mouse embryo. Development, 127: 4455–4467

Hakem, R., Hakem, A., Duncan, G.S., Henderson, J.T., Woo, M., Soengas, M.S., Elia, A., de la Pompa, J.L., Kagi, D., Khoo, W., Potter, J., Yoshida, R., Kaufman, S.A., Lowe, S.W., Penninger, J.M. and Mak, T.W. (1998): Differential requirement for caspase 9 in apoptotic pathways in vivo. Cell, 94: 339–352

Halford, M.M., Armes, J., Buchert, M., Meskenaite, V., Grail, D., Hibbs, M.L., Wilks, A.F., Farlie, P.G., Newgreen, D.F., Hovens, C.M. and Stacker, S.A. (2000): Ryk-deficient mice exhibit craniofacial defects associated with perturbed Eph receptor crosstalk. Nat Genet, 25: 414–418

Hall, B.K. and Miyake, T. (1995): Divide, accumulate, differentiate: cell condensation in skeletal development revisited. Int J Dev Biol, 39: 881–893

Hall, B.K. and Miyake, T. (2000): All for one and one for all: condensations and the initiation of skeletal development. Bioessays, 22: 138–147

Hamburger, V. and Hamilton, H.L. (1951): A series of normal stages in the development of the chick embryo. J. Morphol 88: 49–92

Hammerschmidt, M., Bitgood, M.J. and McMahon, A.P. (1996): Protein kinase A is a common negative regulator of Hedgehog signaling in the vertebrate embryo. Genes Dev, 10: 647–658

Hanken, J. and Thorogood, P. (1993): Evolution and development of the vertebrate skull: the role of pattern formation. Trends Ecol Evol, 8: 9–14.

Hatamura, I., Kanauchi, Y., Takahara, M., Fujiwara, M., Muragaki, Y., Ooshima, A. and Ogino, T. (2001): A nonsense mutation in TRPS1 in a Japanese family with tricho-rhino- phalangeal syndrome type I. Clin Genet, 59: 366–367

Hatta, K., Kimmel, C.B., Ho, R.K. and Walker, C. (1991): The cyclops mutation blocks specification of the floor plate of the zebrafish central nervous system. Nature, 350: 339–341

Hatta, K., Puschel, A.W. and Kimmel, C.B. (1994): Midline signaling in the primordium of the zebrafish anterior central nervous system. Proc Natl Acad Sci U S A, 91: 2061–2065

Hazelton, R.D. (1970): A radioautographic analysis of the migration and fate of cells derived from the occipital somites in the chick embryo with specific reference to the development of the hypoglossal musculature. J Embryol Exp Morphol, 24: 455–466

Heikkinen, J., Toppinen, T., Yeowell, H., Krieg, T., Steinmann, B., Kivirikko, K.I. and Myllyla, R. (1997): Duplication of seven exons in the lysyl hydroxylase gene is associated with longer forms of a repetitive sequence within the gene and is a common cause for the type VI variant of Ehlers-Danlos syndrome. Am J Hum Genet, 60: 48–56

Heisenberg, C.P. and Nusslein-Volhard, C. (1997): The function of silberblick in the positioning of the eye anlage in the zebrafish embryo. Dev Biol, 184: 85–94

Heisenberg, C.P., Tada, M., Rauch, G.J., Saude, L., Concha, M.L., Geisler, R., Stemple, D.L., Smith, J.C. and Wilson, S.W. (2000): Silberblick/Wnt11 mediates convergent extension movements during zebrafish gastrulation. Nature, 405: 76–81

Helbling, P.M., Tran, C.T. and Brandli, A.W. (1998): Requirement for EphA receptor signaling in the segregation of Xenopus third and fourth arch neural crest cells. Mech Dev, 78: 63–79

Helms, J.A., Kim, C.H., Hu, D., Minkoff, R., Thaller, C. and Eichele, G. (1997): Sonic hedgehog participates in craniofacial morphogenesis and is down- regulated by teratogenic doses of retinoic acid. Dev Biol, 187: 25–35

Henderson, D.J., Ybot-Gonzalez, P. and Copp, A.J. (1997): Over-expression of the chondroitin sulphate proteoglycan versican is associated with defective neural crest migration in the Pax3 mutant mouse (splotch). Mech Dev, 69: 39–51

Herrera, E., Samper, E. and Blasco, M.A. (1999): Telomere shortening in mTR-/- embryos is associated with failure to close the neural tube. Embo J, 18: 1172–1181

Heyer, J., Escalante-Alcalde, D., Lia, M., Boettinger, E., Edelmann, W., Stewart, C.L. and Kucherlapati, R. (1999): Postgastrulation Smad2-deficient embryos show defects in embryo turning and anterior morphogenesis. Proc Natl Acad Sci U S A, 96: 12595–12600

Hildebrand, J.D. and Soriano, P. (1999): Shroom, a PDZ domain-containing actin-binding protein, is required for neural tube morphogenesis in mice. Cell, 99: 485–497

Hill, R.E., Favor, J., Hogan, B.L., Ton, C.C., Saunders, G.F., Hanson, I.M., Prosser, J., Jordan, T., Hastie, N.D. and van Heyningen, V. (1991): Mouse small eye results from mutations in a paired-like homeobox- containing gene. Nature, 354: 522–525

Hjalt, T.A., Semina, E.V., Amendt, B.A. and Murray, J.C. (2000): The Pitx2 protein in mouse development. Dev Dyn, 218: 195–200

Hjalt, T.A., Amendt, B.A. and Murray, J.C. (2001): PITX2 regulates procollagen lysyl hydroxylase (PLOD) gene expression: implications for the pathology of Rieger syndrome. J Cell Biol, 152: 545–552

Ho, A.M., Johnson, M.D. and Kingsley, D.M. (2000): Role of the mouse ank gene in control of tissue calcification and arthritis. Science, 289: 265–270

Hofstra, R.M., Osinga, J., Tan-Sindhunata, G., Wu, Y., Kamsteeg, E.J., Stulp, R.P., van Ravenswaaij-Arts, C., Majoor-Krakauer, D., Angrist, M., Chakravarti, A., Meijers, C. and Buys, C.H. (1996): A homozygous mutation in the endothelin-3 gene associated with a combined Waardenburg type 2 and Hirschsprung phenotype (Shah- Waardenburg syndrome). Nat Genet, 12: 445–447

Hoh, J.F. (1992): Muscle fiber types and function. Curr Opin Rheumatol, 4: 801–808

Holst, B.D., Wang, Y., Jones, F.S. and Edelman, G.M. (1997): A binding site for Pax proteins regulates expression of the gene for the neural cell adhesion molecule in the embryonic spinal cord. Proc Natl Acad Sci U S A, 94: 1465–1470

Holtfreter, J. (1968): Mesenchyme and epithelia in inductive and morphological processes. In "Epithelio-mesenchymal Interactions" (P. Fleischmajor and R.B. Billingham, eds.)., pp 1–30. Williams and Williams, Baltimore.

Homanics, G.E., DeLorey, T.M., Firestone, L.L., Quinlan, J.J., Handforth, A., Harrison, N.L., Krasowski, M.D., Rick, C.E., Korpi, E.R., Makela, R., Brilliant, M.H., Hagiwara, N., Ferguson, C., Snyder, K. and Olsen, R.W. (1997): Mice devoid of gamma-aminobutyrate type A receptor beta3 subunit have epilepsy, cleft palate, and hypersensitive behavior. Proc Natl Acad Sci U S A, 94: 4143–4148

Honarpour, N., Du, C., Richardson, J.A., Hammer, R.E., Wang, X. and Herz, J. (2000): Adult Apaf-1-deficient mice exhibit male infertility. Dev Biol, 218: 248–258

Hoover, F. and Glover, J.C. (1998): Regional pattern of retinoid X receptor-alpha gene expression in the central nervous system of the chicken embryo and its up-regulation by exposure to 9-cis retinoic acid. J Comp Neurol, 398: 575–586

Horstadius, G. and Sellman, S. (1946): Experimentalle untersuchungen uber die determination des knorpeligen kopfskelettes bei urodelen. Nova Acta Regiae Soc Sci. Upsaliensis (ser. IV) 13: 1–170

Hosoda, K., Hammer, R.E., Richardson, J.A., Baynash, A.G., Cheung, J.C., Giaid, A. and Yanagisawa, M. (1994): Targeted and natural (piebald-lethal) mutations of endothelin-B receptor gene produce megacolon associated with spotted coat color in mice. Cell, 79: 1267–1276

Hoth, C.F., Milunsky, A., Lipsky, N., Sheffer, R., Clarren, S.K. and Baldwin, C.T. (1993): Mutations in the paired domain of the human PAX3 gene cause Klein- Waardenburg syndrome (WS-III) as well as Waardenburg syndrome type I (WS-I). Am J Hum Genet, 52: 455–462

Houzelstein, D., Cohen, A., Buckingham, M.E. and Robert, B. (1997): Insertional mutation of the mouse Msx1 homeobox gene by an nlacZ reporter gene. Mech Dev, 65: 123–133

Howard, T.D., Paznekas, W.A., Green, E.D., Chiang, L.C., Ma, N., Ortiz de Luna, R.I., Garcia Delgado, C., Gonzalez-Ramos, M., Kline, A.D. and Jabs, E.W. (1997): Mutations in TWIST, a basic helix-loop-helix transcription factor, in Saethre-Chotzen syndrome. Nat Genet, 15: 36–41

Hoving, E.W., Vermeij-Keers, C., Mommaas-Kienhuis, A.M. and Hartwig, N.G. (1990): Separation of neural and surface ectoderm after closure of the rostral neuropore. Anat Embryol, 182: 455–463

Hromas, R., Ye, H., Spinella, M., Dmitrovsky, E., Xu, D., Costa, R.H. (1999): Genesis, a Winged Helix transcriptional repressor, has embryonic expression limited to the neural crest, and stimulates proliferation in vitro in a neural development model. Cell Tissue Res, 297:371–82

Hu, G., Vastardis, H., Bendall, A.J., Wang, Z., Logan, M., Zhang, H., Nelson, C., Stein, S., Greenfield, N., Seidman, C.E., Seidman, J.G. and Abate-Shen, C. (1998): Haploinsufficiency of MSX1: a mechanism for selective tooth agenesis. Mol Cell Biol, 18: 6044–6051

Hu, D. and Helms, J.A. (1999): The role of sonic hedgehog in normal and abnormal craniofacial morphogenesis. Development, 126: 4873–4884

Huang, G.Y., Wessels, A., Smith, B.R., Linask, K.K., Ewart, J.L. and Lo, C.W. (1998): Alteration in connexin 43 gap junction gene dosage impairs conotruncal heart development. Dev Biol, 198: 32–44

Huang, R., Zhi, Q., Izpisua-Belmonte, J.C., Christ, B. and Patel, K. (1999): Origin and development of the avian tongue muscles. Anat Embryol (Berl), 200: 137–152

Huang, R., Zhi, Q., Patel, K., Wilting, J. and Christ, B. (2000): Contribution of single somites to the skeleton and muscles of the occipital and cervical regions in avian embryos. Anat Embryol (Berl), 202: 375–383

Hudson, C.D. and Shapiro, B.L. (1973): A radioautographic study of deoxyribonucleic acid synthesis in embryonic rat palatal shelf epithelium with reference to the concept of programmed cell death. Arch Oral Biol, 18: 77–84

Hudson, R., Taniguchi-Sidle, A., Boras, K., Wiggan, O. and Hamel, P.A. (1998): Alx-4, a transcriptional activator whose expression is restricted to sites of epithelial-mesenchymal interactions. Dev Dyn, 213: 159–169

Hui, C.C. and Joyner, A.L. (1993): A mouse model of greig cephalopolysyndactyly syndrome: the extra-toesJ mutation contains an intragenic deletion of the Gli3 gene. Nat Genet, 3: 241–246

Hunt, P.,Gulisano, M., Cook, M., Sham, M.H., Faiella, A., Wilkinson, D., Boncinelli, E. and Krumlauf, R. (1991): A distinct Hox code for the branchial region of the vertebrate head Nature, 353: 861–4

Hyland, J., Ala-Kokko, L., Royce, P., Steinmann, B., Kivirikko, K.I. and Myllyla, R. (1992): A homozygous stop codon in the lysyl hydroxylase gene in two siblings with Ehlers-Danlos syndrome type VI. Nat Genet, 2: 228–231

Ianakiev, P., Kilpatrick, M.W., Toudjarska, I., Basel, D., Beighton, P. and Tsipouras, P. (2000): Split-hand/split-foot malformation is caused by mutations in the p63 gene on 3q27. Am J Hum Genet, 67: 59–66

Iamaroon, A., Wallon, U.M., Overall, C.M. and Diewert, V.M. (1996): Expression of 72-kDa gelatinase (matrix metalloproteinase-2) in the developing mouse craniofacial complex. Arch Oral Biol, 41: 1109–1119

Iida, K., Koseki, H., Kakinuma, H., Kato, N., Mizutani-Koseki, Y., Ohuchi, H., Yoshioka, H., Noji, S., Kawamura, K., Kataoka, Y., Ueno, F., Taniguchi, M., Yoshida, N., Sugiyama, T. and Miura, N. (1997): Essential roles of the winged helix transcription factor MFH-1 in aortic arch patterning and skeletogenesis. Development, 124: 4627–4638

Ikeya, M., Lee, S.M., Johnson, J.E., McMahon, A.P. and Takada, S. (1997): Wnt signalling required for expansion of neural crest and CNS progenitors. Nature, 389: 966–970

Imai, H., Osumi-Yamashita, N., Ninomiya, Y. and Eto, K. (1996): Contribution of early-emigrating midbrain crest cells to the dental mesenchyme of mandibular molar teeth in rat embryos. Dev Biol, 176: 151–165

Incardona, J.P., Gaffield, W., Kapur, R.P. and Roelink, H. (1998): The teratogenic Veratrum alkaloid cyclopamine inhibits sonic hedgehog signal transduction. Development, 125: 3553–3562

Irving, C. and Mason, I. (2000): Signalling by FGF8 from the isthmus patterns anterior hindbrain and establishes the anterior limit of Hox gene expression. Development, 127, 177–186

Isaac, C., Marsh, K.L., Paznekas, W.A., Dixon, J., Dixon, M.J., Jabs, E.W. and Meier, U.T. (2000): Characterization of the nucleolar gene product, treacle, in Treacher Collins syndrome. Mol Biol Cell, 11: 3061–3071

Ishibashi, M., Ang, S.L., Shiota, K., Nakanishi, S., Kageyama, R. and Guillemot, F. (1995): Targeted disruption of mammalian hairy and Enhancer of split homolog-1 (HES-1) leads to up-regulation of neural helix-loop-helix factors, premature neurogenesis, and severe neural tube defects. Genes Dev, 9: 3136–3148

Itasaki, N., Sharpe, J., Morrison, A. and Krumlauf, R. (1996): Reprogramming Hox expression in the vertebrate hindbrain: influence of paraxial mesoderm and rhombomere transposition. Neuron, 16: 487–500

Itoh, T., Ikeda, T., Gomi, H., Nakao, S., Suzuki, T. and Itohara, S. (1997): Unaltered secretion of beta-amyloid precursor protein in gelatinase A (matrix metalloproteinase 2)-deficient mice. J Biol Chem, 272: 22389–22392

Iulianella, A. and Lohnes, D. (1997): Contribution of retinoic acid receptor gamma to retinoid-induced craniofacial and axial defects. Dev Dyn, 209: 92–104

Jacobson, A.G. (1988): Somitomeres: mesodermal segments of vertebrate embryos. Development, 104: 209–220

Jaskoll, T., Greenberg, G., Melnick, M. (1991): Neural tube and neural crest: a new view with time-lapse high-definition photomicroscopy. Am J Med Genet, 41:333-45

Jeffs, P., Jaques, K. and Osmond, M. (1992): Cell death in cranial neural crest development. Anat Embryol, 185: 583–588

Jernvall, J. and Thesleff, I. (2000): Reiterative signaling and patterning during mammalian tooth morphogenesis. Mech Dev, 92: 19–29

Jerome, L.A. and Papaioannou, V.E. (2001): DiGeorge syndrome phenotype in mice mutant for the T-box gene, Tbx1. Nat Genet, 27: 286–291

Jiang, R., Lan, Y., Chapman, H.D., Shawber, C., Norton, C.R., Serreze, D.V., Weinmaster, G. and Gridley, T. (1998a): Defects in limb, craniofacial, and thymic development in Jagged2 mutant mice. Genes Dev, 12: 1046–1057

Jiang, R., Lan, Y., Norton, C.R., Sundberg, J.P. and Gridley, T. (1998b): The Slug gene is not essential for mesoderm or neural crest development in mice. Dev Biol, 198: 277–285

Jiang, X., Rowitch, D.H., Soriano, P., McMahon, A.P. and Sucov, H.M. (2000): Fate of the mammalian cardiac neural crest. Development, 127: 1607–1616

Jiang, X., Iseki, S., Maxson, R.E., Sucov, H.M. and Morriss-Kay, G.M. (2002): Tissue origins and interactions in the Mammalian skull vault. Dev Biol, 241: 106–116

Johnston, M.C. (1966): A radioautographic study of the migration and fate of cranial neural crest cells in the chick embryo. Anat Rec, 156: 143–155

Johnston, M.C. and Bronsky, P.T. (1991): Animal models for human craniofacial malformations. J Craniofac Genet Dev Biol, 11: 277–291

Jones, K.L., Smith, D.W., Harvey, M.A., Hall, B.D. and Quan, L. (1975): Older paternal age and fresh gene mutation: data on additional disorders. J Pediatr, 86: 84–88

Jones, T.L., Karavanova, I., Maeno, M., Ong, R.C., Kung, H.F. and Daar, I.O. (1995): Expression of an amphibian homolog of the Eph family of receptor tyrosine kinases is developmentally regulated. Oncogene, 10: 1111–1117

Jones, N.C., Farlie, P.G., Minichiello, J. and Newgreen, D.F. (1999): Detection of an appropriate kinase activity in branchial arches I and II that coincides with peak expression of the Treacher Collins syndrome gene product, treacle. Hum Mol Genet, 8: 2239–2245

Jorgenson, R.J., Levin, L.S., Cross, H.E., Yoder, F. and Kelly, T.E. (1978): The Rieger syndrome. Am J Med Genet, 2: 307–318

Jung, H.S., Francis-West, P.H., Widelitz, R.B., Jiang, T.X., Ting-Berreth, S., Tickle, C., Wolpert, L. and Chuong, C.M. (1998): Local inhibitory action of BMPs and their relationships with activators in feather formation: implications for periodic patterning. Dev Biol, 196: 11–23

Juriloff, D.M., Harris, M.J., Tom, C. and MacDonald, K.B. (1991): Normal mouse strains differ in the site of initiation of closure of the cranial neural tube. Teratology, 44: 225–233

Kaartinen, V., Voncken, J.W., Shuler, C., Warburton, D., Bu, D., Heisterkamp, N. and Groffen, J. (1995): Abnormal lung development and cleft palate in mice lacking TGF-beta 3 indicates defects of epithelial-mesenchymal interaction. Nat Genet, 11: 415–421

Juriloff, D.M., Harris, M.J. (2000): Mouse models for neural tube closure defects. Hum Mol Genet, 9:993–1000

Kaartinen, V., Cui, X.M., Heisterkamp, N., Groffen, J. and Shuler, C.F. (1997): Transforming growth factor-beta3 regulates transdifferentiation of medial edge epithelium during palatal fusion and associated degradation of the basement membrane. Dev Dyn, 209: 255–260

Kalff-Suske, M., Wild, A., Topp, J., Wessling, M., Jacobsen, E.M., Bornholdt, D., Engel, H., Heuer, H., Aalfs, C.M., Ausems, M.G., Barone, R., Herzog, A., Heutink, P., Homfray, T., Gillessen-Kaesbach,

G., Konig, R., Kunze, J., Meinecke, P., Muller, D., Rizzo, R., Strenge, S., Superti-Furga, A. and Grzeschik, K.H. (1999): Point mutations throughout the GLI3 gene cause Greig cephalopolysyndactyly syndrome. Hum Mol Genet, 8: 1769–1777

Kalter, H. (1992): Cyclopia produced in a very early retinoic acid experiment. Teratology, 46: 207–208

Kambe, M., Miyamoto, Y. and Hayashi, M. (1998): Characterization of the integrin alpha(v) gene promoter in mice. Biochim Biophys Acta, 1395: 209–219

Kanzler, B., Kuschert, S.J., Liu, Y.H. and Mallo, M. (1998): Hoxa-2 restricts the chondrogenic domain and inhibits bone formation during development of the branchial area. Development, 125: 2587–2597

Kanzler, B., Foreman, R.K., Labosky, P.A., Mallo, M. (2000): BMP signaling is essential for development of skeletogenic and neurogenic cranial neural crest. Development, 127:1095-104

Kasarskis, A., Manova, K. and Anderson, K.V. (1998): A phenotype-based screen for embryonic lethal mutations in the mouse. Proc Natl Acad Sci U S A, 95: 7485–7490

Kastner, P., Mark, M., Ghyselinck, N., Krezel, W., Dupe, V., Grondona, J.M. and Chambon, P. (1997): Genetic evidence that the retinoid signal is transduced by heterodimeric RXR/RAR functional units during mouse development. Development, 124: 313–326

Kaufman, M.H., Chang, H.H. and Shaw, J.P. (1995): Craniofacial abnormalities in homozygous Small eye (Sey/Sey) embryos and newborn mice. J Anat, 186: 607–617

Kazanskaya, O., Glinka, A. and Niehrs, C. (2000): The role of Xenopus dickkopf1 in prechordal plate specification and neural patterning. Development, 127: 4981–4992

Kelley, R.I. (1998): RSH/Smith-Lemli-Opitz syndrome: mutations and metabolic morphogenesis. Am J Hum Genet, 63: 322–326

Kempf, H., Linares, C., Corvol, P. and Gasc, J.M. (1998): Pharmacological inactivation of the endothelin type A receptor in the early chick embryo: a model of mispatterning of the branchial arch derivatives. Development, 125: 4931–4941

Kessel, M. and Gruss, P. (1991): Homeotic transformations of murine vertebrae and concomitant alteration of Hox codes induced by retinoic acid. Cell, 67: 89–104

Kibar, Z., Vogan, K.J., Groulx, N., Justice, M.J., Underhill, D.A. and Gros, P. (2001): Ltap, a mammalian homolog of Drosophila Strabismus/Van Gogh, is altered in the mouse neural tube mutant Looptail. Nat Genet, 28: 251–255

Kiecker, C., Muller, F., Wu, W., Glinka, A., Strahle, U. and Niehrs, C. (2000): Phenotypic effects in Xenopus and zebrafish suggest that one-eyed pinhead functions as antagonist of BMP signalling. Mech Dev, 94: 37–46

Kiecker, C. and Niehrs, C. (2001): The role of prechordal mesendoderm in neural patterning. Curr Opin Neurobiol, 11: 27–33

Kil, S.H., Krull, C.E., Cann, G., Clegg, D. and Bronner-Fraser, M. (1998): The alpha4 subunit of integrin is important for neural crest cell migration. Dev Biol, 202: 29–42

Kimber, W.L., Hsieh, P., Hirotsune, S., Yuva-Paylor, L., Sutherland, H.F., Chen, A., Ruiz-Lozano, P., Hoogstraten-Miller, S.L., Chien, K.R., Paylor, R., Scambler, P.J. and Wynshaw-Boris, A. (1999): Deletion of 150 kb in the minimal DiGeorge/velocardiofacial syndrome critical region in mouse. Hum Mol Genet, 8: 2229–2237

Kimmel, C.B., Miller, C.T. and Keynes, R.J. (2001): Neural crest patterning and the evolution of the jaw. J Anat, 199: 105–120

Kingsley, D.M., Bland, A.E., Grubber, J.M., Marker, P.C., Russell, L.B., Copeland, N.G. and Jenkins, N.A. (1992): The mouse short ear skeletal morphogenesis locus is associated with defects in a bone morphogenetic member of the TGF beta superfamily. Cell, 71: 399–410

Kirby, M.L., Gale, T.F. and Stewart, D.E. (1983): Neural crest cells contribute to normal aorticopulmonary septation. Science, 220: 1059–1061

Kirby, M.L. and Waldo, K.L. (1990): Role of neural crest in congenital heart disease. Circulation, 82: 332–340

Kirby, M.L. and Waldo, K.L. (1995): Neural crest and cardiovascular patterning. Circ Res, 77: 211–215

Klootwijk, R., Franke, B., van der Zee, C.E., de Boer, R.T., Wilms, W., Hol, F.A. and Mariman, E.C. (2000): A deletion encompassing Zic3 in bent tail, a mouse model for X-linked neural tube defects. Hum Mol Genet, 9: 1615–1622

Koleske, A.J., Gifford, A.M., Scott, M.L., Nee, M., Bronson, R.T., Miczek, K.A. and Baltimore, D. (1998): Essential roles for the Abl and Arg tyrosine kinases in neurulation. Neuron, 21: 1259–1272

Köntges, G. and Lumsden, A. (1996): Rhombencephalic neural crest segmentation is preserved throughout craniofacial ontogeny. Development, 122: 3229–3242

Korfage, J.A., Schueler, Y.T., Brugman, P. and Van Eijden, T.M. (2001): Differences in myosin heavy-chain composition between human jaw-closing muscles and supra- and infrahyoid muscles. Arch Oral Biol, 46: 821–827

Kosaka, K. and Eto, K. (1986): Appearance of a unique cell type in the fusion sites of facial processes. J Craniofac Genet Dev Biol Suppl, 2: 45–52

Kozlowski, K. and Walter, M.A. (2000): Variation in residual PITX2 activity underlies the phenotypic spectrum of anterior segment developmental disorders. Hum Mol Genet, 9: 2131–2139

Krauss, S., Concordet, J.P. and Ingham, P.W. (1993): A functionally conserved homolog of the Drosophila segment polarity gene hh is expressed in tissues with polarizing activity in zebrafish embryos. Cell, 75: 1431–1444

Krotoski, D.M., Domingo, C. and Bronner-Fraser, M. (1986): Distribution of a putative cell surface receptor for fibronectin and laminin in the avian embryo. J Cell Biol, 103: 1061–1071

Krull, C.E. (2001): Segmental organization of neural crest migration. Mech Dev, 105: 37–45

Kuan, C.Y., Yang, D.D., Samanta Roy, D.R., Davis, R.J., Rakic, P. and Flavell, R.A. (1999): The Jnk1 and Jnk2 protein kinases are required for regional specific apoptosis during early brain development. Neuron, 22: 667–676

Kubota, Y. and Ito, K. (2000): Chemotactic migration of mesencephalic neural crest cells in the mouse. Dev Dyn, 217: 170–179

Kuida, K., Zheng, T.S., Na, S., Kuan, C., Yang, D., Karasuyama, H., Rakic, P. and Flavell, R.A. (1996): Decreased apoptosis in the brain and premature lethality in CPP32- deficient mice. Nature, 384: 368–372

Kuida, K., Haydar, T.F., Kuan, C.Y., Gu, Y., Taya, C., Karasuyama, H., Su, M.S., Rakic, P. and Flavell, R.A. (1998): Reduced apoptosis and cytochrome c-mediated caspase activation in mice lacking caspase 9. Cell, 94: 325–337

Kulak, S.C., Kozlowski, K., Semina, E.V., Pearce, W.G. and Walter, M.A. (1998): Mutation in the RIEG1 gene in patients with iridogoniodysgenesis syndrome. Hum Mol Genet, 7: 1113–1117

Kulesa, P.M. and Fraser, S.E. (1998): Neural crest cell dynamics revealed by time-lapse video microscopy of whole embryo chick explant cultures. Dev Biol, 204: 327–344

Kunte, A., Ivey, K., Yamagishi, C., Garg, V., Yamagishi, H. and Srivastava, D. (2001): A common cis-acting sequence in the DiGeorge critical region regulates bi-directional transcription of UFD1L and CDC45L. Mech Dev, 108: 81–92

Kuo, J.S., Patel, M., Gamse, J., Merzdorf, C., Liu, X., Apekin, V. and Sive, H. (1998): Opl: a zinc finger protein that regulates neural determination and patterning in Xenopus. Development, 125: 2867–2882

Kurahashi, H., Tsuda, E., Kohama, R., Nakayama, T., Masuno, M., Imaizumi, K., Kamiya, T., Sano, T., Okada, S. and Nishisho, I. (1997): Another critical region for deletion of 22q11: a study of 100 patients. Am J Med Genet, 72: 180–185

Kuratani, S.C. and Eichele, G. (1993): Rhombomere transplantation repatterns the segmental organization of cranial nerves and reveals cell-autonomous expression of a homeodomain protein. Development, 117: 105–117

Kuratani, S., Nobusada, Y., Horigome, N., Shigetani, Y. (2001): Embryology of the lamprey and evolution of the vertebrate jaw: insights from molecular and developmental perspectives. Phil Trans Roy Soc 356: 1615–1632.

Kurihara, Y., Kurihara, H., Suzuki, H., Kodama, T., Maemura, K., Nagai, R., Oda, H., Kuwaki, T., Cao, W.H., Kamada, N. and et al. (1994): Elevated blood pressure and craniofacial abnormalities in mice deficient in endothelin-1. Nature, 368: 703–710

Kurihara, Y., Kurihara, H., Oda, H., Maemura, K., Nagai, R., Ishikawa, T. and Yazaki, Y. (1995): Aortic arch malformations and ventricular septal defect in mice deficient in endothelin-1. J Clin Invest, 96: 293–300

LaBonne, C. and Bronner-Fraser, M. (1999): Molecular mechanisms of neural crest formation. Annu Rev Cell Dev Biol, 15: 81–112

LaBonne, C. and Bronner-Fraser, M. (2000): Snail-related transcriptional repressors are required in Xenopus for both the induction of the neural crest and its subsequent migration. Dev Biol, 221: 195–205

Ladher, R.K., Anakwe, K.U., Gurney, A.L., Schoenwolf, G.C., Francis-West, P.H. (2000a): Identification of synergistic signals initiating inner ear development.
Science, 290:1965–7
Ladher, R.K., Church, V.L., Allen, S., Robson, L., Abdelfattah, A., Brown, N.A., Hattersley, G., Rosen, V., Luyten, F.P., Dale, L. and Francis-West, P.H. (2000): Cloning and expression of the Wnt antagonists Sfrp-2 and Frzb during chick development. Dev Biol, 218: 183–198
Lallier, T. and Bronner-Fraser, M. (1992): Alpha 1 beta 1 integrin on neural crest cells recognizes some laminin substrata in a Ca(2+)-independent manner. J Cell Biol, 119: 1335–1345
Lallier, T., Leblanc, G., Artinger, K.B. and Bronner-Fraser, M. (1992): Cranial and trunk neural crest cells use different mechanisms for attachment to extracellular matrices. Development, 116: 531–541
Lallier, T. and Bronner-Fraser, M. (1993): Inhibition of neural crest cell attachment by integrin antisense oligonucleotides. Science, 259: 692–695
Lammer, E.J. and Opitz, J.M. (1986): The DiGeorge anomaly as a developmental field defect. Am J Med Genet Suppl, 2: 113–127
Lamour, V., Lecluse, Y., Desmaze, C., Spector, M., Bodescot, M., Aurias, A., Osley, M.A. and Lipinski, M. (1995): A human homolog of the S. cerevisiae HIR1 and HIR2 transcriptional repressors cloned from the DiGeorge syndrome critical region. Hum Mol Genet, 4: 791–799
Lanctot, C., Moreau, A., Chamberland, M., Tremblay, M.L. and Drouin, J. (1999): Hindlimb patterning and mandible development require the Ptx1 gene. Development, 126: 1805–1810
Lanier, L.M., Gates, M.A., Witke, W., Menzies, A.S., Wehman, A.M., Macklis, J.D., Kwiatkowski, D., Soriano, P. and Gertler, F.B. (1999): Mena is required for neurulation and commissure formation. Neuron, 22: 313–325
Lauder, J.M. and Zimmerman, E.F. (1988): Sites of serotonin uptake in epithelia of the developing mouse palate, oral cavity, and face: possible role in morphogenesis. J Craniofac Genet Dev Biol, 8: 265–276
Laverty, H.G. and Wilson, J.B. (1998): Murine CASK is disrupted in a sex-linked cleft palate mouse mutant. Genomics, 53: 29–41
Lavrin, I.O., McLean, W., Seegmiller, R.E., Olsen, B.R. and Hay, E.D. (2001): The mechanism of palatal clefting in the Col11a1 mutant mouse. Arch Oral Biol, 46: 865–869
Lawson, K.A., Meneses, J.J. and Pedersen, R.A. (1991): Clonal analysis of epiblast fate during germ layer formation in the mouse embryo. Development, 113: 891–911
Lawson, K.A., Dunn, N.R., Roelen, B.A., Zeinstra, L.M., Davis, A.M., Wright, C.V., Korving, J.P. and Hogan, B.L. (1999): Bmp4 is required for the generation of primordial germ cells in the mouse embryo. Genes Dev, 13: 424–436
Lecanda, F., Warlow, P.M., Sheikh, S., Furlan, F., Steinberg, T.H. and Civitelli, R. (2000): Connexin43 deficiency causes delayed ossification, craniofacial abnormalities, and osteoblast dysfunction. J Cell Biol, 151: 931–944
Le Douarin, N.M. and Teillet, M.A. (1974): Experimental analysis of the migration and differentiation of neuroblasts of the autonomic nervous system and of neurectodermal mesenchymal derivatives, using a biological cell marking technique. Dev Biol, 41: 162–184
Le Douarin, N.M., Jotereau, F.V., Houssaint, E. and Belo, M. (1976): Ontogeny of the avian thymus and bursa of Fabricius studied in interspecific chimeras. Ann Immunol (Paris), 127: 849–856
Le Douarin, N.M., Teillet, M.A. and Le Lievre, C. (1977): Influence of the tissue environment on the differentiation of neural crest cells. Soc Gen Physiol Ser, 32: 11–27
Le Douarin, N.M., Ziller, C. and Couly, G.F. (1993): Patterning of neural crest derivatives in the avian embryo: in vivo and in vitro studies. Dev Biol, 159: 24–49
Lee, Y.M., Osumi-Yamashita, N., Ninomiya, Y., Moon, C.K., Eriksson, U. and Eto, K. (1995): Retinoic acid stage-dependently alters the migration pattern and identity of hindbrain neural crest cells. Development, 121: 825–837
Lee, S-H., Fu, K. K., Hui. J.N. and Richman, J.M. (2001); Noggin and retinoic acid transform the identity of avian facial prominences. Nature, 414: 909–912
Le Lièvre, C.S. and Le Douarin, N.M. (1975): Mesenchymal derivatives of the neural crest: analysis of chimaeric quail and chick embryos. J Embryol Exp Morphol, 34: 125–154
Le Lièvre, C.S. (1978): Participation of neural crest-derived cells in the genesis of the skull in birds. J Embryol Exp Morphol, 47: 17–37

Le Lièvre, C.S., Schweizer, G.G., Ziller, C.M. and Le Douarin, N.M. (1980): Restrictions of developmental capabilities in neural crest cell derivatives as tested by in vivo transplantation experiments. Dev Biol, 77: 362–378

Lemire, R.J. (1988): Neural tube defects. Jama, 259: 558–562

Lettice, L.A., Purdie, L.A., Carlson, G.J., Kilanowski, F., Dorin, J. and Hill, R.E. (1999): The mouse bagpipe gene controls development of axial skeleton, skull, and spleen. Proc Natl Acad Sci U S A, 96: 9695–9700

Levy, A., Demczuk, S., Aurias, A., Depetris, D., Mattei, M.G. and Philip, N. (1995): Interstitial 22q11 microdeletion excluding the ADU breakpoint in a patient with DiGeorge syndrome. Hum Mol Genet, 4: 2417–2419

Li, H.S., Yang, J.M., Jacobson, R.D., Pasko, D. and Sundin, O. (1994): Pax-6 is first expressed in a region of ectoderm anterior to the early neural plate: implications for stepwise determination of the lens. Dev Biol, 162: 181–194

Li, H., Tierney, C., Wen, L., Wu, J.Y. and Rao, Y. (1997a): A single morphogenetic field gives rise to two retina primordia under the influence of the prechordal plate. Development, 124: 603–615

Li, Q.Y., Newbury-Ecob, R.A., Terrett, J.A., Wilson, D.I., Curtis, A.R., Yi, C.H., Gebuhr, T., Bullen, P.J., Robson, S.C., Strachan, T., Bonnet, D., Lyonnet, S., Young, I.D., Raeburn, J.A., Buckler, A.J., Law, D.J. and Brook, J.D. (1997b): Holt-Oram syndrome is caused by mutations in TBX5, a member of the Brachyury (T) gene family. Nat Genet, 15: 21–29

Li, Q., Lu, Q., Hwang, J.Y., Büscher, D., Lee, K-F., Izpisua-Belmonte, J.C. and Verma, I.M. (1999): IKK1-deficient mice exhibit abnotmal development of skin and skeleton. Genes Dev, 13: 1322–28

Lin, C.R., Kioussi, C., O'Connell, S., Briata, P., Szeto, D., Liu, F., Izpisua-Belmonte, J.C. and Rosenfeld, M.G. (1999): Pitx2 regulates lung asymmetry, cardiac positioning and pituitary and tooth morphogenesis. Nature, 401: 279–282

Lindsay, E.A., Botta, A., Jurecic, V., Carattini-Rivera, S., Cheah, Y.C., Rosenblatt, H.M., Bradley, A. and Baldini, A. (1999): Congenital heart disease in mice deficient for the DiGeorge syndrome region. Nature, 401: 379–383

Lindsay, E.A., Vitelli, F., Su, H., Morishima, M., Huynh, T., Pramparo, T., Jurecic, V., Ogunrinu, G., Sutherland, H.F., Scambler, P.J., Bradley, A., Baldini, A. (2001): Tbx1 haploinsufficieny in the DiGeorge syndrome region causes aortic arch defects in mice. Nature, 410:97–101

Liu, J.P. and Jessell, T.M. (1998): A role for rhoB in the delamination of neural crest cells from the dorsal neural tube. Development, 125: 5055–5067

Liu, J., Prickett, T.D., Elliott, E., Meroni, G. and Brautigan, D.L. (2001): Phosphorylation and microtubule association of the Opitz syndrome protein mid-1 is regulated by protein phosphatase 2A via binding to the regulatory subunit alpha 4. Proc Natl Acad Sci U S A, 98: 6650–6655

Lo, C.W., Cohen, M.F., Huang, G.Y., Lazatin, B.O., Patel, N., Sullivan, R., Pauken, C. and Park, S.M. (1997): Cx43 gap junction gene expression and gap junctional communication in mouse neural crest cells. Dev Genet, 20: 119–132

Logan, M., Pagan-Westphal, S.M., Smith, D.M., Paganessi, L. and Tabin, C.J. (1998): The transcription factor Pitx2 mediates situs-specific morphogenesis in response to left-right asymmetric signals. Cell, 94: 307–317

Lohnes, D., Mark, M., Mendelsohn, C., Dolle, P., Dierich, A., Gorry, P., Gansmuller, A. and Chambon, P. (1994): Function of the retinoic acid receptors (RARs) during development (I). Craniofacial and skeletal abnormalities in RAR double mutants. Development, 120: 2723–2748

Lorain, S., Quivy, J.P., Monier-Gavelle, F., Scamps, C., Lecluse, Y., Almouzni, G. and Lipinski, M. (1998): Core histones and HIRIP3, a novel histone-binding protein, directly interact with WD repeat protein HIRA. Mol Cell Biol, 18: 5546–5556

Loring, J.F., Barker, D.L. and Erickson, C.A. (1988): Migration and differentiation of neural crest and ventral neural tube cells in vitro: Implications for in vitro and in vivo studies of the neural crest. J. Neurosci, 8: 1001–1015

Lu, M.F., Cheng, H.T., Kern, M.J., Potter, S.S., Tran, B., Diekwisch, T.G. and Martin, J.F. (1999a): Prx-1 functions cooperatively with another paired-related homeobox gene, prx-2, to maintain cell fates within the craniofacial mesenchyme. Development, 126: 495–504

Lu, M.F., Pressman, C., Dyer, R., Johnson, R.L. and Martin, J.F. (1999b): Function of Rieger syndrome gene in left-right asymmetry and craniofacial development. Nature, 401: 276–278

Lunn, E.R., Scourfield, J., Keynes, R.J. and Stern, C.D. (1987): The neural tube origin of ventral root sheath cells in the chick embryo. Development, 101: 247–54

Lumsden, A.G. (1988): Spatial organization of the epithelium and the role of neural crest cells in the initiation of the mammalian tooth germ. Development, 103: 155–169

Lumsden, A. and Keynes, R. (1989): Segmental patterns of neuronal development in the chick hindbrain. Nature, 337: 424–428

Lumsden, A., Sprawson, N. and Graham, A. (1991): Segmental origin and migration of neural crest cells in the hindbrain region of the chick embryo. Development, 113: 1281–1291

Luo, G., Hofmann, C., Bronckers, A.L., Sohocki, M., Bradley, A. and Karsenty, G. (1995): BMP-7 is an inducer of nephrogenesis, and is also required for eye development and skeletal patterning. Genes Dev, 9: 2808–2820

Luo, J., Sucov, H.M., Bader, J.A., Evans, R.M. and Giguere, V. (1996): Compound mutants for retinoic acid receptor (RAR) beta and RAR alpha 1 reveal developmental functions for multiple RAR beta isoforms. Mech Dev, 55: 33–44

Luo, B., Aster, J.C., Hasserjian, R.P., Kuo, F. and Sklar, J. (1997): Isolation and functional analysis of a cDNA for human Jagged2, a gene encoding a ligand for the Notch1 receptor. Mol Cell Biol, 17: 6057–6067

Lyons, G.E., Micales, B.K., Herr, M.J., Horrigan, S.K., Namciu, S., Shardy, D., Stavnezer, E. (1994): Protooncogene c-ski is expressed in both proliferating and postmitotic neuronal populations. Dev Dyn, 201:354-65

MacArthur, C.A., Lawshe, A., Xu, J., Santos-Ocampo, S., Heikinheimo, M., Chellaiah, A.T. and Ornitz, D.M. (1995): FGF-8 isoforms activate receptor splice forms that are expressed in mesenchymal regions of mouse development. Development, 121: 3603–3613

Macdonald, K.B., Juriloff, D.M. and Harris, M.J. (1989): Developmental study of neural tube closure in a mouse stock with a high incidence of exencephaly. Teratology, 39: 195–213

Macdonald, R., Barth, K.A., Xu, Q., Holder, N., Mikkola, I. and Wilson, S.W. (1995): Midline signalling is required for Pax gene regulation and patterning of the eyes. Development, 121: 3267–3278

Mackenzie, S., Walsh, F.S. and Graham, A. (1998): Migration of hypoglossal myoblast precursors. Dev Dyn, 213: 349–358

Maden, M. (1999): Heads or tails? Retinoic acid will decide. Bioessays, 21: 809–812

Maemura, K., Kurihara, H., Kurihara, Y., Oda, H., Ishikawa, T., Copeland, N.G., Gilbert, D.J., Jenkins, N.A. and Yazaki, Y. (1996): Sequence analysis, chromosomal location, and developmental expression of the mouse preproendothelin-1 gene. Genomics, 31: 177–184

Magnaghi, P., Roberts, C., Lorain, S., Lipinski, M. and Scambler, P.J. (1998): HIRA, a mammalian homologue of Saccharomyces cerevisiae transcriptional co-repressors, interacts with Pax3. Nat Genet, 20: 74–77

Mahmood, R., Mason, I.J. and Morriss-Kay, G.M. (1996): Expression of Fgf-3 in relation to hindbrain segmentation, otic pit position and pharyngeal arch morphology in normal and retinoic acid-exposed mouse embryos. Anat Embryol (Berl), 194: 13–22

Makino, S., Masuya, H., Ishijima, J., Yada, Y. and Shiroishi, T. (2001): A Spontaneous Mouse Mutation, mesenchymal dysplasia (mes), Is Caused by a Deletion of the Most C-Terminal Cytoplasmic Domain of patched (ptc). Dev Biol, 239: 95–106

Mammi, I., De Giorgio, P., Clementi, M. and Tenconi, R. (1998): Cardiovascular anomaly in Rieger Syndrome: heterogeneity or contiguity? Acta Ophthalmol Scand, 76: 509–512

Mancilla, A. and Mayor, R. (1996): Neural crest formation in Xenopus laevis: mechanisms of Xslug induction. Dev Biol, 177: 580–589

Mansour, S.L., Goddard, J.M. and Capecchi, M.R. (1993): Mice homozygous for a targeted disruption of the proto-oncogene int-2 have developmental defects in the tail and inner ear. Development, 117: 13–28

Marchant, L., Linker, C., Ruiz, P., Guerrero, N. and Mayor, R. (1998): The inductive properties of mesoderm suggest that the neural crest cells are specified by a BMP gradient. Dev Biol, 198: 319–329

Marcucio, R.S. and Noden, D.M. (1999): Myotube heterogeneity in developing chick craniofacial skeletal muscles. Dev Dyn, 214: 178–194

Mariani, F.V. and Harland, R.M. (1998): XBF-2 is a transcriptional repressor that converts ectoderm into neural tissue. Development, 125: 5019–5031

Marin, F. and Charnay, P. (2000): Hindbrain patterning: FGFs regulate Krox20 and mafB/kr expression in the otic/preotic region. Development, 127: 4925–4935

Marlow, F., Zwartkruis, F., Malicki, J., Neuhauss, S.C., Abbas, L., Weaver, M., Driever, W. and Solnica-Krezel, L. (1998): Functional interactions of genes mediating convergent extension, knypek and trilobite, during the partitioning of the eye primordium in zebrafish. Dev Biol, 203: 382–399

Marsh, K.L., Dixon, J. and Dixon, M.J. (1998): Mutations in the Treacher Collins syndrome gene lead to mislocalization of the nucleolar protein treacle. Hum Mol Genet, 7: 1795–1800

Marshall, H., Nonchev, S., Sham, M.H., Muchamore, I., Lumsden, A. and Krumlauf, R. (1992): Retinoic acid alters hindbrain Hox code and induces transformation of rhombomeres 2/3 into a 4/5 identity. Nature, 360: 737–741

Marti, E., Takada, R., Bumcrot, D.A., Sasaki, H. and McMahon, A.P. (1995): Distribution of Sonic hedgehog peptides in the developing chick and mouse embryo. Development, 121: 2537–2547

Martínez-Alvarez, C., Bonelli, R., Tudela, C., Gato, A., Mena, J., O'Kane, S. and Ferguson, M.W. (2000a): Bulging medial edge epithelial cells and palatal fusion. Int J Dev Biol, 44: 331-335

Martinez-Alvarez, C., Tudela, C., Perez-Miguelsanz, J., O'Kane, S., Puerta, J. and Ferguson, M.W. (2000b): Medial edge epithelial cell fate during palatal fusion. Dev Biol, 220: 343–357

Martinsen, B.J. and Bronner-Fraser, M. (1998): Neural crest specification regulated by the helix-loop-helix repressor Id2. Science, 281: 988–991

Mason, I.J., Fuller-Pace, F., Smith, R. and Dickson, C. (1994): FGF-7 (keratinocyte growth factor) expression during mouse development suggests roles in myogenesis, forebrain regionalisation and epithelial- mesenchymal interactions. Mech Dev, 45: 15–30

Matsuo, T., Osumi-Yamashita, N., Noji, S., Ohuchi, H., Koyama, E., Myokai, F., Matsuo, N., Taniguchi, S., Doi, H., Iseki, S. and et al. (1993): A mutation in the Pax-6 gene in rat small eye is associated with impaired migration of midbrain crest cells. Nat Genet, 3: 299–304

Matzuk, M.M., Kumar, T.R. and Bradley, A. (1995a): Different phenotypes for mice deficient in either activins or activin receptor type II. Nature, 374: 356–360

Matzuk, M.M., Kumar, T.R., Vassalli, A., Bickenbach, J.R., Roop, D.R., Jaenisch, R. and Bradley, A. (1995b): Functional analysis of activins during mammalian development. Nature, 374: 354–356

Matzuk, M.M., Lu, N., Vogel, H., Sellheyer, K., Roop, D.R. and Bradley, A. (1995c): Multiple defects and perinatal death in mice deficient in follistatin. Nature, 374: 360–363

Mavrogiannis, L.A., Antonopoulou, I., Baxova, A., Kutilek, S., Kim, C.A., Sugayama, S.M., Salamanca, A., Wall, S.A., Morriss-Kay, G.M. and Wilkie, A.O. (2001): Haploinsufficiency of the human homeobox gene ALX4 causes skull ossification defects. Nat Genet, 27: 17–18

McClearn, D. and Noden, D.M. (1988): Ontogeny of architectural complexity in embryonic quail visceral arch muscles. Am J Anat, 183: 277–293

McGonnell, I.M., McKay, I.J. and Graham, A. (2001): A population of caudally migrating cranial neural crest cells: functional and evolutionary implications. Dev Biol, 236: 354–363

Meier, S.P. (1982): The development of segmentation in the cranial region of vertebrate embryos. Scan Electron Microsc, (Pt: 1269–1282

Meier, S. and Tam, P.P. (1982): Metameric pattern development in the embryonic axis of the mouse. I. Differentiation of the cranial segments. Differentiation, 21: 95–108

Meier, U.T. and Blobel, G. (1992): Nopp140 shuttles on tracks between nucleolus and cytoplasm. Cell, 70: 127–138

Melhuish, T.A. and Wotton, D. (2000): The interaction of the carboxyl terminus-binding protein with the Smad corepressor TGIF is disrupted by a holoprosencephaly mutation in TGIF. J Biol Chem, 275: 39762–39766

Melkoniemi, M., Brunner, H.G., Manouvrier, S., Hennekam, R., Superti-Furga, A., Kaariainen, H., Pauli, R.M., van Essen, T., Warman, M.L., Bonaventure, J., Miny, P. and Ala-Kokko, L. (2000): Autosomal recessive disorder otospondylomegaepiphyseal dysplasia is associated with loss-of-function mutations in the COL11A2 gene. Am J Hum Genet, 66: 368–377

Mendelsohn, C., Lohnes, D., Decimo, D., Lufkin, T., LeMeur, M., Chambon, P. and Mark, M. (1994a): Function of the retinoic acid receptors (RARs) during development (II). Multiple abnormalities at various stages of organogenesis in RAR double mutants. Development, 120: 2749–2771

Mendelsohn, C., Mark, M., Dolle, P., Dierich, A., Gaub, M.P., Krust, A., Lampron, C. and Chambon, P. (1994b): Retinoic acid receptor beta 2 (RAR beta 2) null mutant mice appear normal. Dev Biol, 166: 246–258

Meno, C., Gritsman, K., Ohishi, S., Ohfuji, Y., Heckscher, E., Mochida, K., Shimono, A., Kondoh, H., Talbot, W.S., Robertson, E.J., Schier, A.F. and Hamada, H. (1999): Mouse Lefty2 and zebrafish antivin are feedback inhibitors of nodal signaling during vertebrate gastrulation. Mol Cell, 4: 287–298

Meredith, J.E., Jr., Fazeli, B. and Schwartz, M.A. (1993): The extracellular matrix as a cell survival factor. Mol Biol Cell, 4: 953–961

Merino, R., Rodriguez-Leon, J., Macias, D., Ganan, Y., Economides, A.N. and Hurle, J.M. (1999): The BMP antagonist Gremlin regulates outgrowth, chondrogenesis and programmed cell death in the developing limb. Development, 126: 5515–5522

Merscher, S., Funke, B., Epstein, J.A., Heyer, J., Puech, A., Lu, M.M., Xavier, R.J., Demay, M.B., Russell, R.G., Factor, S., Tokooya, K., Jore, B.S., Lopez, M., Pandita, R.K., Lia, M., Carrion, D., Xu, H., Schorle, H., Kobler, J.B., Scambler, P., Wynshaw-Boris, A., Skoultchi, A.I., Morrow, B.E. and Kucherlapati, R. (2001): TBX1 is responsible for cardiovascular defects in velo-cardio- facial/DiGeorge syndrome. Cell, 104: 619–629

Miau, L.H., Chang, C.J., Tsai, W.H. and Lee, S.C. (1997): Identification and characterization of a nucleolar phosphoprotein, Nopp140, as a transcription factor. Mol Cell Biol, 17: 230–239

Miettinen, P.J., Chin, J.R., Shum, L., Slavkin, H.C., Shuler, C.F., Derynck, R. and Werb, Z. (1999): Epidermal growth factor receptor function is necessary for normal craniofacial development and palate closure. Nat Genet, 22: 69–73

Miller, R.P. and Becker, B.A. (1975): Teratogenicity of oral diazepam and diphenylhydantoin in mice. Toxicol Appl Pharmacol, 32: 53–61

Miller, C.T., Schilling, T.F., Lee, K., Parker, J. and Kimmel, C.B. (2000): Sucker encodes a zebrafish Endothelin-1 required for ventral pharyngeal arch development. Development, 127: 3815–3828

Mills, A.A., Zheng, B., Wang, X.J., Vogel, H., Roop, D.R. and Bradley, A. (1999): p63 is a p53 homologue required for limb and epidermal morphogenesis. Nature, 398: 708–713

Mina, M. and Kollar, E.J. (1987): The induction of odontogenesis in non-dental mesenchyme combined with early murine mandibular arch epithelium. Arch Oral Biol, 32: 123–127

Miner, J.H., Cunningham, J. and Sanes, J.R. (1998): Roles for laminin in embryogenesis: exencephaly, syndactyly, and placentopathy in mice lacking the laminin alpha5 chain. J Cell Biol, 143: 1713–1723

Ming, J.E., Roessler, E. and Muenke, M. (1998): Human developmental disorders and the Sonic hedgehog pathway. Mol Med Today, 4: 343–349

Mitsiadis, T.A., Mucchielli, M.L., Raffo, S., Proust, J.P., Koopman, P. and Goridis, C. (1998): Expression of the transcription factors Otlx2, Barx1 and Sox9 during mouse odontogenesis. Eur J Oral Sci, 106 Suppl 1: 112–116

Mizugishi, K., Aruga, J., Nakata, K. and Mikoshiba, K. (2001): Molecular properties of Zic proteins as transcriptional regulators and their relationship to GLI proteins. J Biol Chem, 276: 2180–2188

Mo, R., Freer, A.M., Zinyk, D.L., Crackower, M.A., Michaud, J., Heng, H.H., Chik, K.W., Shi, X.M., Tsui, L.C., Cheng, S.H., Joyner, A.L. and Hui, C. (1997): Specific and redundant functions of Gli2 and Gli3 zinc finger genes in skeletal patterning and development. Development, 124: 113–123

Mogass, M., Bringas, P., Jr. and Shuler, C.F. (2000): Characterization of desmosomal component expression during palatogenesis. Int J Dev Biol, 44: 317–322

Moiseiwitsch, J.R. and Lauder, J.M. (1995): Serotonin regulates mouse cranial neural crest migration. Proc Natl Acad Sci U S A, 92: 7182–7186

Momeni, P., Glockner, G., Schmidt, O., von Holtum, D., Albrecht, B., Gillessen-Kaesbach, G., Hennekam, R., Meinecke, P., Zabel, B., Rosenthal, A., Horsthemke, B. and Ludecke, H.J. (2000): Mutations in a new gene, encoding a zinc-finger protein, cause tricho- rhino-phalangeal syndrome type I. Nat Genet, 24: 71–74

Monsoro-Burq, A.H., Duprez, D., Watanabe, Y., Bontoux, M., Vincent, C., Brickell, P. and Le Douarin, N. (1996): The role of bone morphogenetic proteins in vertebral development. Development, 122: 3607–3616

Mootoosamy, R.C. and Dietrich, S (2002): Distinct regulatory cascades for head and trunk myogenesis. Development 129: 573–583

Morasso, M.I., Grinberg, A., Robinson, G., Sargent, T.D. and Mahon, K.A. (1999): Placental failure in mice lacking the homeobox gene Dlx3. Proc Natl Acad Sci U S A, 96: 162–167
Moro Balbas, J.A., Gato, A., Alonso Revuelta, M.I., Pastor, J.F., Repressa, J.J. and Barbosa, E. (1993): Retinoic acid induces changes in the rhombencephalic neural crest cells migration and extracellular matrix composition in chick embryos. Teratology, 48: 197–206
Morriss-Kay, G. and Tuckett, F. (1985): The role of microfilaments in cranial neurulation in rat embryos: effects of short-term exposure to cytochalasin D. J Embryol Exp Morphol, 88: 333–348
Morriss-Kay, G.M., Murphy, P., Hill, R.E. and Davidson, D.R. (1991): Effects of retinoic acid excess on expression of Hox-2.9 and Krox-20 and on morphological segmentation in the hindbrain of mouse embryos. Embo J, 10: 2985–2995
Morriss-Kay, G., Ruberte, E. and Fukiishi, Y. (1993): Mammalian neural crest and neural crest derivatives. Anat Anz, 175: 501–507
Morriss-Kay, G.M., Iseki, S. and Johnson, D. (2001): Genetic control of the cell proliferation-differentiation balance in the developing skull vault: roles of fibroblast growth factor receptor signalling pathways. Novartis Found Symp, 232: 102–116
Morrison-Graham, K., Schatteman, G.C., Bork, T., Bowen-Pope, D.F. and Weston, J.A. (1992): A PDGF receptor mutation in the mouse (Patch) perturbs the development of a non-neuronal subset of neural crest-derived cells. Development, 115: 133–142
Moury, J.D. and Jacobson, A.G. (1990): The origins of neural crest cells in the axolotl. Dev Biol, 141: 243–253
Muccielli, M.L., Martinez, S., Pattyn, A., Goridis, C. and Brunet, J.F. (1996): Otlx2, an Otx-related homeobox gene expressed in the pituitary gland and in a restricted pattern in the forebrain. Mol Cell Neurosci, 8: 258–271
Muhr, J., Jessell, T.M. and Edlund, T. (1997): Assignment of early caudal identity to neural plate cells by a signal from caudal paraxial mesoderm. Neuron, 19: 487–502
Muhr, J., Graziano, E., Wilson, S., Jessell, T.M. and Edlund, T. (1999): Convergent inductive signals specify midbrain, hindbrain, and spinal cord identity in gastrula stage chick embryos. Neuron, 23: 689–702
Müller, F., Albert, S., Blader, P., Fischer, N., Hallonet, M. and Strahle, U. (2000): Direct action of the nodal-related signal cyclops in induction of sonic hedgehog in the ventral midline of the CNS. Development, 127: 3889–3897
Mundlos, S., Otto, F., Mundlos, C., Mulliken, J.B., Aylsworth, A.S., Albright, S., Lindhout, D., Cole, W.G., Henn, W., Knoll, J.H., Owen, M.J., Mertelsmann, R., Zabel, B.U. and Olsen, B.R. (1997): Mutations involving the transcription factor CBFA1 cause cleidocranial dysplasia. Cell, 89: 773–779
Munoz-Sanjuan, I., Cooper, M.K., Beachy, P.A., Fallon, J.F. and Nathans, J. (2001): Expression and regulation of chicken fibroblast growth factor homologous factor (FHF)-4 during craniofacial morphogenesis. Dev Dyn, 220: 238–245
Nagai, T., Aruga, J., Takada, S., Gunther, T., Sporle, R., Schughart, K. and Mikoshiba, K. (1997): The expression of the mouse Zic1, Zic2, and Zic3 gene suggests an essential role for Zic genes in body pattern formation. Dev Biol, 182: 299–313
Nagai, T., Aruga, J., Minowa, O., Sugimoto, T., Ohno, Y., Noda, T., Mikoshiba, K. (2000): Zic2 regulates the kinetics of neurulation. Proc Natl Acad Sci U S A. 97:1618–23
Nakagawa, S. and Takeichi, M. (1995): Neural crest cell-cell adhesion controlled by sequential and subpopulation-specific expression of novel cadherins. Development, 121: 1321–1332
Nakagawa, S. and Takeichi, M. (1998): Neural crest emigration from the neural tube depends on regulated cadherin expression. Development, 125: 2963–2971
Nakamura, H. and Ayer-le Lievre, C.S. (1982): Mesectodermal capabilities of the trunk neural crest of birds. J Embryol Exp Morphol, 70: 1–18
Nakata, K., Nagai, T., Aruga, J. and Mikoshiba, K. (1997): Xenopus Zic3, a primary regulator both in neural and neural crest development. Proc Natl Acad Sci U S A, 94: 11980–11985
Nakata, K., Nagai, T., Aruga, J. and Mikoshiba, K. (1998): Xenopus Zic family and its role in neural and neural crest development. Mech Dev, 75: 43–51
Nakata, K., Koyabu, Y., Aruga, J. and Mikoshiba, K. (2000): A novel member of the Xenopus Zic family, Zic5, mediates neural crest development. Mech Dev, 99: 83–91
Nanni, L., Ming, J.E., Bocian, M., Steinhaus, K., Bianchi, D.W., Die-Smulders, C., Giannotti, A., Imaizumi, K., Jones, K.L., Campo, M.D., Martin, R.A., Meinecke, P., Pierpont, M.E., Robin, N.H.,

Young, I.D., Roessler, E. and Muenke, M. (1999): The mutational spectrum of the sonic hedgehog gene in holoprosencephaly: SHH mutations cause a significant proportion of autosomal dominant holoprosencephaly. Hum Mol Genet, 8: 2479–2488
Nanni, L., Ming, J.E., Du, Y., Hall, R.K., Aldred, M., Bankier, A. and Muenke, M. (2001): SHH mutation is associated with solitary median maxillary central incisor: a study of 13 patients and review of the literature. Am J Med Genet, 102: 1–10
Nataf, V., Amemiya, A., Yanagisawa, M. and Le Douarin, N.M. (1998): The expression pattern of endothelin 3 in the avian embryo. Mech Dev, 73: 217–220
Nellemann, C., de Bellard, M.E., Barembaum, M., Laufer, E. and Bronner-Fraser, M. (2001): Excess lunatic fringe causes cranial neural crest over-proliferation. Dev Biol, 235: 121–130
Neubüser, A., Peters, H., Balling, R. and Martin, G.R. (1997): Antagonistic interactions between FGF and BMP signaling pathways: a mechanism for positioning the sites of tooth formation. Cell, 90: 247–255
Newgreen, D. and Thiery, J.P. (1980): Fibronectin in early avian embryos: synthesis and distribution along the migration pathways of neural crest cells. Cell Tissue Res, 211: 269–291
Nguyen, V.H., Schmid, B., Trout, J., Connors, S.A., Ekker, M. and Mullins, M.C. (1998): Ventral and lateral regions of the zebrafish gastrula, including the neural crest progenitors, are established by a Bmp2b/swirl pathway of genes. Dev Biol, 199: 93–110
Nichols, D.H. (1981): Neural crest formation in the head of the mouse embryo as observed using a new histological technique. J Embryol Exp Morphol, 64:105–20
Nichols, D.H. (1986): Formation and distribution of neural crest mesenchyme to the first pharyngeal arch region of the mouse embryo. Am J Anat, 176: 221–231
Niederlander, C. and Lumsden, A. (1996): Late emigrating neural crest cells migrate specifically to the exit points of cranial branchiomotor nerves. Development, 122: 2367–2374
Niederreither, K., Vermot, J., Schuhbaur, B., Chambon, P. and Dolle, P. (2000): Retinoic acid synthesis and hindbrain patterning in the mouse embryo. Development, 127: 75–85
Nieto, M.A. (2001): The early steps of neural crest development. Mech Dev, 105: 27–35
Nieto, M.A., Sargent, M.G., Wilkinson, D.G. and Cooke, J. (1994): Control of cell behavior during vertebrate development by Slug, a zinc finger gene. Science, 264: 835–839
Nieto, M.A., Sechrist, J., Wilkinson, D.G. and Bronner-Fraser, M. (1995): Relationship between spatially restricted Krox-20 gene expression in branchial neural crest and segmentation in the chick embryo hindbrain. Embo J, 14: 1697–1710
Nishibatake, M., Kirby, M.L. and Van Mierop, L.H. (1987): Pathogenesis of persistent truncus arteriosus and dextroposed aorta in the chick embryo after neural crest ablation. Circulation, 75: 255–264
Niswander, L. and Martin, G.R. (1993): FGF-4 and BMP-2 have opposite effects on limb growth. Nature, 361: 68–71
Noden, D.M. (1975): An analysis of migratory behavior of avian cephalic neural crest cells. Dev Biol, 42: 106–130
Noden, D.M. (1978a): The control of avian cephalic neural crest cytodifferentiation. I. Skeletal and connective tissues. Dev Biol, 67: 296–312
Noden, D.M. (1978b): The control of avian cephalic neural crest cytodifferentiation. II. Neural tissues. Dev Biol, 67: 313–329
Noden, D.M. (1982): Patterns and organization of craniofacial skeletogenic and myogenic mesenchyme: a perspective. Prog Clin Biol Res, 101: 167–203
Noden, D.M. (1983a): The embryonic origins of avian cephalic and cervical muscles and associated connective tissues. Am J Anat, 168: 257–276
Noden, D.M. (1983b): The role of the neural crest in patterning of avian cranial skeletal, connective, and muscle tissues. Dev Biol, 96: 144–165
Noden, D.M. (1984): Craniofacial development: new views on old problems. Anat Rec, 208: 1–13
Noden, D.M. (1986): Origins and patterning of craniofacial mesenchymal tissues. J Craniofac Genet Dev Biol Suppl, 2: 15–31
Noden, D.M. (1988): Interactions and fates of avian craniofacial mesenchyme. Development, 103: 121–140
Noden, D.M. (1991a): Cell movements and control of patterned tissue assembly during craniofacial development. J Craniofac Genet Dev Biol, 11: 192–213

Noden, D.M. (1991b): Origins and patterning of avian outflow tract endocardium. Development, 111: 867–876

Noden, D.M., Marcucio, R., Borycki, A.G. and Emerson, C.P., Jr. (1999): Differentiation of avian craniofacial muscles: I. Patterns of early regulatory gene expression and myosin heavy chain synthesis. Dev Dyn, 216: 96–112

Nomura, M. and Li, E. (1998): Smad2 role in mesoderm formation, left-right patterning and craniofacial development. Nature, 393: 786–790

Noramly, S. and Morgan, B.A. (1998): BMPs mediate lateral inhibition at successive stages in feather tract development. Development, 125: 3775–3787

Nottoli, T., Hagopian-Donaldson, S., Zhang, J., Perkins, A. and Williams, T. (1998): AP-2-null cells disrupt morphogenesis of the eye, face, and limbs in chimeric mice. Proc Natl Acad Sci U S A, 95: 13714–13719

Nugent, P., Sucov, H.M., Pisano, M.M. and Greene, R.M. (1999): The role of RXR-alpha in retinoic acid-induced cleft palate as assessed with the RXR-alpha knockout mouse. Int J Dev Biol, 43: 567–570

Nürnberg, P., Thiele, H., Chandler, D., Hohne, W., Cunningham, M.L., Ritter, H., Leschik, G., Uhlmann, K., Mischung, C., Harrop, K., Goldblatt, J., Borochowitz, Z.U., Kotzot, D., Westermann, F., Mundlos, S., Braun, H.S., Laing, N. and Tinschert, S. (2001): Heterozygous mutations in ANKH, the human ortholog of the mouse progressive ankylosis gene, result in craniometaphyseal dysplasia. Nat Genet, 28: 37–41

Oda, T., Elkahloun, A.G., Pike, B.L., Okajima, K., Krantz, I.D., Genin, A., Piccoli, D.A., Meltzer, P.S., Spinner, N.B., Collins, F.S. and Chandrasekharappa, S.C. (1997): Mutations in the human Jagged1 gene are responsible for Alagille syndrome. Nat Genet, 16: 235–242

Odent, S., Atti-Bitach, T., Blayau, M., Mathieu, M., Aug, J., Delezo de, A.L., Gall, J.Y., Le Marec, B., Munnich, A., David, V. and Vekemans, M. (1999): Expression of the Sonic hedgehog (SHH) gene during early human development and phenotypic expression of new mutations causing holoprosencephaly. Hum Mol Genet, 8: 1683–1689

Opperman, L.A., Sweeney, T.M., Redmon, J., Persing, J.A. and Ogle, R.C. (1993): Tissue interactions with underlying dura mater inhibit osseous obliteration of developing cranial sutures. Dev Dyn, 198: 312–322

Opperman, L.A., Persing, J.A., Sheen, R. and Ogle, R.C. (1994): In the absence of periosteum, transplanted fetal and neonatal rat coronal sutures resist osseous obliteration. J Craniofac Surg, 5: 327–332

Opperman, L.A., Passarelli, R.W., Morgan, E.P., Reintjes, M. and Ogle, R.C. (1995): Cranial sutures require tissue interactions with dura mater to resist osseous obliteration in vitro. J Bone Miner Res, 10: 1978–1987

Opperman, L.A., Chhabra, A., Nolen, A.A., Bao, Y. and Ogle, R.C. (1998): Dura mater maintains rat cranial sutures in vitro by regulating suture cell proliferation and collagen production. J Craniofac Genet Dev Biol, 18: 150–158

Opperman, L.A. (2000): Cranial sutures as intramembranous bone growth sites. Dev Dyn, 219: 472–485

Orioli, D., Henkemeyer, M., Lemke, G., Klein, R. and Pawson, T. (1996): Sek4 and Nuk receptors cooperate in guidance of commissural axons and in palate formation. Embo J, 15: 6035–6049

Orrico, A., Galli, L., Falciani, M., Bracci, M., Cavaliere, M.L., Rinaldi, M.M., Musacchio, A. and Sorrentino, V. (2000): A mutation in the pleckstrin homology (PH) domain of the FGD1 gene in an Italian family with faciogenital dysplasia (Aarskog-Scott syndrome). FEBS Lett, 478: 216–220

Osumi-Yamashita, N., Ninomiya, Y., Doi, H. and Eto, K. (1994): The contribution of both forebrain and midbrain crest cells to the mesenchyme in the frontonasal mass of mouse embryos. Dev Biol, 164: 409–419

Osumi-Yamashita, N., Ninomiya Y., Doi, H., and Eto, K. (1996): Rhombomere formation and hindbrain crest cell migration from prorhombomeric origins in mouse embryos. Develop. Growth Differ., 38: 107–118

Osumi-Yamashita, N., Kuratani, S., Ninomiya, Y., Aoki, K., Iseki, S., Chareonvit, S., Doi, H., Fujiwara, M., Watanabe, T. and Eto, K. (1997): Cranial anomaly of homozygous rSey rat is associated with a defect in the migration pathway of midbrain crest cells. Dev Growth Differ, 39: 53–67

Pace, J.M., Li, Y., Seegmiller, R.E., Teuscher, C., Taylor, B.A. and Olsen, B.R. (1997): Disproportionate micromelia (Dmm) in mice caused by a mutation in the C- propeptide coding region of Col2a1. Dev Dyn, 208: 25–33

Pardanaud, L., Altmann, C., Kitos, P., Dieterlen-Lievre, F. and Buck, C.A. (1987): Vasculogenesis in the early quail blastodisc as studied with a monoclonal antibody recognizing endothelial cells. Development, 100: 339–349

Pasqualetti, M., Ori, M., Nardi, I. and Rijli, F.M. (2000): Ectopic Hoxa2 induction after neural crest migration results in homeosis of jaw elements in Xenopus. Development, 127: 5367–5378

Pasteris, N.G., Cadle, A., Logie, L.J., Porteous, M.E., Schwartz, C.E., Stevenson, R.E., Glover, T.W., Wilroy, R.S. and Gorski, J.L. (1994): Isolation and characterization of the faciogenital dysplasia (Aarskog- Scott syndrome) gene: a putative Rho/Rac guanine nucleotide exchange factor. Cell, 79: 669–678

Paznekas, W.A., Zhang, N., Gridley, T. and Jabs, E.W. (1997): Mouse TCOF1 is expressed widely, has motifs conserved in nucleolar phosphoproteins, and maps to chromosome 18. Biochem Biophys Res Commun, 238: 1–6

Peault, B.M., Thiery, J.P. and Le Douarin, N.M. (1983): Surface marker for hemopoietic and endothelial cell lineages in quail that is defined by a monoclonal antibody. Proc Natl Acad Sci U S A, 80: 2976–2980

Pelton, R.W., Hogan, B.L., Miller, D.A. and Moses, H.L. (1990): Differential expression of genes encoding TGFs beta 1, beta 2, and beta 3 during murine palate formation. Dev Biol, 141: 456–460

Pera, E.M. and Kessel, M. (1997): Patterning of the chick forebrain anlage by the prechordal plate. Development, 124: 4153–4162

Pera, E., Stein, S. and Kessel, M. (1999): Ectodermal patterning in the avian embryo: epidermis versus neural plate. Development, 126: 63–73

Pera, E.M. and De Robertis, E.M. (2000): A direct screen for secreted proteins in Xenopus embryos identifies distinct activities for the Wnt antagonists Crescent and Frzb-1. Mech Dev, 96: 183–195

Perris, R. (1997): The extracellular matrix in neural crest-cell migration. Trends Neurosci, 20: 23–31

Peters, H., Neubuser, A., Kratochwil, K. and Balling, R. (1998): Pax9-deficient mice lack pharyngeal pouch derivatives and teeth and exhibit craniofacial and limb abnormalities. Genes Dev, 12: 2735–2747

Piccolo, S., Agius, E., Leyns, L., Bhattacharyya, S., Grunz, H., Bouwmeester, T. and De Robertis, E.M. (1999): The head inducer Cerberus is a multifunctional antagonist of Nodal, BMP and Wnt signals. Nature, 397: 707–710

Pilia, G., Uda, M., Macis, D., Frau, F., Crisponi, L., Balli, F., Barbera, C., Colombo, C., Frediani, T., Gatti, R., Iorio, R., Marazzi, M.G., Marcellini, M., Musumeci, S., Nebbia, G., Vajro, P., Ruffa, G., Zancan, L., Cao, A. and DeVirgilis, S. (1999): Jagged-1 mutation analysis in Italian Alagille syndrome patients. Hum Mutat, 14: 394–400

Pizette, S. and Niswander, L. (1999): BMPs negatively regulate structure and function of the limb apical ectodermal ridge. Development, 126: 883–894

Poelmann, R.E. and Gittenberger-de Groot, A.C. (1999): A subpopulation of apoptosis-prone cardiac neural crest cells targets to the venous pole: multiple functions in heart development? Dev Biol, 207: 271–286

Pohl, B.S. and Knöchel, W. (2001): Overexpression of the transcriptional repressor FoxD3 prevents neural crest formation in Xenopus embryos. Mech Dev, 103: 93–106

Poole, T.J. and Coffin, J.D. (1988): Developmental angiogenesis: quail embryonic vasculature. Scanning Microsc, 2: 443–448

Poole, T.J. and Thiery, J.P. (1986): Antibodies and a synthetic peptide that block cell-fibronectin adhesion arrest neural crest cell migration in vivo. Prog Clin Biol Res, 235–238

Porter, J.D., Khanna, S., Kaminski, H.J., Rao, J.S., Merriam, A.P., Richmonds, C.R., Leahy, P., Li, J. and Andrade, F.H. (2001): Extraocular muscle is defined by a fundamentally distinct gene expression profile. Proc Natl Acad Sci U S A, 98: 12062–12067

Posnick, J.C., Bortoluzzi, P. and Armstrong, D.C. (1994): Nasal dermoid sinus cysts: an unusual presentation, computed tomographic scan findings, and surgical results. Ann Plast Surg, 32: 519–523

Pratt, R.M. and Hassell, J.R. (1975): Appearance and distribution of carbohydrate-rich macromolecules on the epithelial surface of the developing rat palatal shelf. Dev Biol, 45: 192–198

Pratt, R.M. and Martin, G.R. (1975): Epithelial cell death and cyclic AMP increase during palatal development. Proc Natl Acad Sci U S A, 72: 874–877

Price, J.A., Bowden, D.W., Wright, J.T., Pettenati, M.J. and Hart, T.C. (1998): Identification of a mutation in DLX3 associated with tricho-dento- osseous (TDO) syndrome. Hum Mol Genet, 7: 563–569

Prince, V. and Lumsden, A. (1994): Hoxa-2 expression in normal and transposed rhombomeres: independent regulation in the neural tube and neural crest. Development, 120:911–23

Proetzel, G., Pawlowski, S.A., Wiles, M.V., Yin, M., Boivin, G.P., Howles, P.N., Ding, J., Ferguson, M.W. and Doetschman, T. (1995): Transforming growth factor-beta 3 is required for secondary palate fusion. Nat Genet, 11: 409–414

Puech, A., Saint-Jore, B., Merscher, S., Russell, R.G., Cherif, D., Sirotkin, H., Xu, H., Factor, S., Kucherlapati, R. and Skoultchi, A.I. (2000): Normal cardiovascular development in mice deficient for 16 genes in 550 kb of the velocardiofacial/DiGeorge syndrome region. Proc Natl Acad Sci U S A, 97: 10090–10095

Puffenberger, E.G., Hosoda, K., Washington, S.S., Nakao, K., deWit, D., Yanagisawa, M. and Chakravart, A. (1994): A missense mutation of the endothelin-B receptor gene in multigenic Hirschsprung's disease. Cell, 79: 1257–1266

Puschel, A.W., Gruss, P. and Westerfield, M. (1992): Sequence and expression pattern of pax-6 are highly conserved between zebrafish and mice. Development, 114: 643–651

Qiu, M., Bulfone, A., Martinez, S., Meneses, J.J., Shimamura, K., Pedersen, R.A. and Rubenstein, J.L. (1995): Null mutation of Dlx-2 results in abnormal morphogenesis of proximal first and second branchial arch derivatives and abnormal differentiation in the forebrain. Genes Dev, 9: 2523–2538

Qiu, M., Bulfone, A., Ghattas, I., Meneses, J.J., Christensen, L., Sharpe, P.T., Presley, R., Pedersen, R.A. and Rubenstein, J.L. (1997): Role of the Dlx homeobox genes in proximodistal patterning of the branchial arches: mutations of Dlx-1, Dlx-2, and Dlx-1 and -2 alter morphogenesis of proximal skeletal and soft tissue structures derived from the first and second arches. Dev Biol, 185: 165–184

Qu, S., Niswender, K.D., Ji, Q., van der Meer, R., Keeney, D., Magnuson, M.A. and Wisdom, R. (1997): Polydactyly and ectopic ZPA formation in Alx-4 mutant mice. Development, 124: 3999–4008

Qu, S., Tucker, S.C., Zhao, Q., deCrombrugghe, B. and Wisdom, R. (1999): Physical and genetic interactions between Alx4 and Cart1. Development, 126: 359–369

Quaderi, N.A., Schweiger, S., Gaudenz, K., Franco, B., Rugarli, E.I., Berger, W., Feldman, G.J., Volta, M., Andolfi, G., Gilgenkrantz, S., Marion, R.W., Hennekam, R.C., Opitz, J.M., Muenke, M., Ropers, H.H. and Ballabio, A. (1997): Opitz G/BBB syndrome, a defect of midline development, is due to mutations in a new RING finger gene on Xp22. Nat Genet, 17: 285–291

Raven, C.P. and Kloos, J. (1945): Induction by medial and lateral pieces of the archenteron roof with special reference to the determination of the neural crest. Acta Néerl Morph, 5: 348–362.

Rebagliati, M.R., Toyama, R., Haffter, P. and Dawid, I.B. (1998): Cyclops encodes a nodal-related factor involved in midline signaling. Proc Natl Acad Sci U S A, 95: 9932–9937

Revel, J.P. and Brown, S.S. (1976): Cell junctions in development, with particular reference to the neural tube. Cold Spring Harb Symp Quant Biol, 40: 443–455

Richman, J.M. and Tickle, C. (1989): Epithelia are interchangeable between facial primordia of chick embryos and morphogenesis is controlled by the mesenchyme. Dev Biol, 136: 201–210

Richman, J.M. and Tickle, C. (1992): Epithelial-mesenchymal interactions in the outgrowth of limb buds and facial primordia in chick embryos. Dev Biol, 154: 299–308

Richman, J.M., Herbert, M., Matovinovic, E. and Walin, J. (1997): Effect of fibroblast growth factors on outgrowth of facial mesenchyme. Dev Biol, 189: 135–147

Rijli, F.M., Mark, M., Lakkaraju, S., Dierich, A., Dolle, P. and Chambon, P. (1993): A homeotic transformation is generated in the rostral branchial region of the head by disruption of Hoxa-2, which acts as a selector gene. Cell, 75: 1333–1349

Rivera-Pérez, J.A., Mallo, M., Gendron-Maguire, M., Gridley, T. and Behringer, R.R. (1995): Goosecoid is not an essential component of the mouse gastrula organizer but is required for craniofacial and rib development. Development, 121: 3005–3012

Rivera-Perez, J.A., Wakamiya, M. and Behringer, R.R. (1999): Goosecoid acts cell autonomously in mesenchyme-derived tissues during craniofacial development. Development, 126: 3811–3821

Robbins, J.R., McGuire, P.G., Wehrle-Haller, B. and Rogers, S.L. (1999): Diminished matrix metalloproteinase 2 (MMP-2) in ectomesenchyme-derived tissues of the Patch mutant mouse: regulation of MMP-2 by PDGF and effects on mesenchymal cell migration. Dev Biol, 212: 255–263

Roberts, C., Daw, S.C., Halford, S. and Scambler, P.J. (1997): Cloning and developmental expression analysis of chick Hira (Chira), a candidate gene for DiGeorge syndrome. Hum Mol Genet, 6: 237–245

Robinson, V., Smith, A., Flenniken, A.M. and Wilkinson, D.G. (1997): Roles of Eph receptors and ephrins in neural crest pathfinding. Cell Tissue Res, 290: 265–274

Robson, L.G. (1993): Cellular patterning of fast and slow fibres in the intermandibularis muscle of chick embryos. Development, 117: 329–339

Roessler, E., Belloni, E., Gaudenz, K., Jay, P., Berta, P., Scherer, S.W., Tsui, L.C. and Muenke, M. (1996): Mutations in the human Sonic Hedgehog gene cause holoprosencephaly. Nat Genet, 14: 357–360

Roessler, E., Belloni, E., Gaudenz, K., Vargas, F., Scherer, S.W., Tsui, L.C. and Muenke, M. (1997): Mutations in the C-terminal domain of Sonic Hedgehog cause holoprosencephaly. Hum Mol Genet, 6: 1847–1853

Roessler, E. and Muenke, M. (2001): Midline and laterality defects: left and right meet in the middle. Bioessays, 23: 888–900

Rosenquist, G.C. (1981): Epiblast origin and early migration of neural crest cells in the chick embryo. Dev Biol, 87: 201–211

Rowe, A., Eager, N.S. and Brickell, P.M. (1991): A member of the RXR nuclear receptor family is expressed in neural- crest-derived cells of the developing chick peripheral nervous system. Development, 111: 771–778

Rowe, A., Richman, J.M. and Brickell, P.M. (1992): Development of the spatial pattern of retinoic acid receptor-beta transcripts in embryonic chick facial primordia. Development, 114: 805–813

Rowe, A. and Brickell, P.M. (1995): Expression of the chicken retinoid X receptor-gamma gene in migrating cranial neural crest cells. Anat Embryol (Berl), 192: 1–8

Ruland, J., Duncan, G.S., Elia, A., del Barco Barrantes, I., Nguyen, L., Plyte, S., Millar, D.G., Bouchard, D., Wakeham, A., Ohashi, P.S. and Mak, T.W. (2001): Bcl10 is a positive regulator of antigen receptor-induced activation of NF-kappaB and neural tube closure. Cell, 104: 33–42

Runyan, R.B., Maxwell, G.D. and Shur, B.D. (1986): Evidence for a novel enzymatic mechanism of neural crest cell migration on extracellular glycoconjugate matrices. J Cell Biol, 102: 432–441

Ryan, A.K., Blumberg, B., Rodriguez-Esteban, C., Yonei-Tamura, S., Tamura, K., Tsukui, T., de la Pena, J., Sabbagh, W., Greenwald, J., Choe, S., Norris, D.P., Robertson, E.J., Evans, R.M., Rosenfeld, M.G. and Izpisua Belmonte, J.C. (1998): Pitx2 determines left-right asymmetry of internal organs in vertebrates. Nature, 394: 545–551

Saadi, I., Semina, E.V., Amendt, B.A., Harris, D.J., Murphy, K.P., Murray, J.C. and Russo, A.F. (2001): Identification of a dominant negative homeodomain mutation in Rieger syndrome. J Biol Chem, 276: 23034–23041

Sadaghiani, B. and Thiebaud, C.H. (1987): Neural crest development in the Xenopus laevis embryo, studied by interspecific transplantation and scanning electron microscopy. Dev Biol, 124: 91–110

Sadeghi-Nejad, A. and Senior, B. (1974): Autosomal dominant transmission of isolated growth hormone deficiency in iris-dental dysplasia (Rieger's syndrome). J Pediatr, 85: 644–648

Sah, V.P., Attardi, L.D., Mulligan, G.J., Williams, B.O., Bronson, R.T. and Jacks, T. (1995): A subset of p53-deficient embryos exhibit exencephaly. Nat Genet, 10: 175–180

Sakai, Y. (1989): Neurulation in the mouse: manner and timing of neural tube closure. Anat Rec, 223: 194–203

Saldivar, J.R., Krull, C.E., Krumlauf, R., Ariza-McNaughton, L. and Bronner-Fraser, M. (1996): Rhombomere of origin determines autonomous versus environmentally regulated expression of Hoxa-3 in the avian embryo. Development, 122: 895–904

Saldivar, J.R., Sechrist, J.W., Krull, C.E., Ruffins, S. and Bronner-Fraser, M. (1997): Dorsal hindbrain ablation results in rerouting of neural crest migration and changes in gene expression, but normal hyoid development. Development, 124: 2729–2739

Salvarezza, S.B. and Rovasio, R.A. (1997): Exogenous retinoic acid decreases in vivo and in vitro proliferative activity during the early migratory stage of neural crest cells. Cell Prolif, 30: 71–80

Sampath, K., Rubinstein, A.L., Cheng, A.M., Liang, J.O., Fekany, K., Solnica-Krezel, L., Korzh, V., Halpern, M.E. and Wright, C.V. (1998): Induction of the zebrafish ventral brain and floorplate requires cyclops/nodal signalling. Nature, 395: 185–189
Sanford, L.P., Ormsby, I., Gittenberger-de Groot, A.C., Sariola, H., Friedman, R., Boivin, G.P., Cardell, E.L. and Doetschman, T. (1997): TGFbeta2 knockout mice have multiple developmental defects that are non- overlapping with other TGFbeta knockout phenotypes. Development, 124: 2659–2670
Sarkar, L., Cobourne, M., Naylor, S., Smalley, M., Dale, T. and Sharpe, P.T. (2000): Wnt/Shh interactions regulate ectodermal boundary formation during mammalian tooth development. Proc Natl Acad Sci U S A, 97: 4520–4524
Sarkar, S., Petiot, A., Copp, A., Ferretti, P. and Thorogood, P. (2001): FGF2 promotes skeletogenic differentiation of cranial neural crest cells. Development, 128: 2143–2152
Satoda, M., Zhao, F., Diaz, G.A., Burn, J., Goodship, J., Davidson, H.R., Pierpont, M.E. and Gelb, B.D. (2000): Mutations in TFAP2B cause Char syndrome, a familial form of patent ductus arteriosus. Nat Genet, 25: 42–46
Satokata, I. and Maas, R. (1994): Msx1 deficient mice exhibit cleft palate and abnormalities of craniofacial and tooth development. Nat Genet, 6: 348–356
Satokata, I., Ma, L., Ohshima, H., Bei, M., Woo, I., Nishizawa, K., Maeda, T., Takano, Y., Uchiyama, M., Heaney, S., Peters, H., Tang, Z., Maxson, R. and Maas, R. (2000): Msx2 deficiency in mice causes pleiotropic defects in bone growth and ectodermal organ formation. Nat Genet, 24: 391–395
Saúde, L., Woolley, K., Martin, P., Driever, W. and Stemple, D.L. (2000): Axis-inducing activities and cell fates of the zebrafish organizer. Development, 127: 3407–3417
Saxton, T.M., Ciruna, B.G., Holmyard, D., Kulkarni, S., Harpal, K., Rossant, J. and Pawson, T. (2000): The SH2 tyrosine phosphatase shp2 is required for mammalian limb development. Nat Genet, 24: 420–423.
Schatteman, G.C., Morrison-Graham, K., van Koppen, A., Weston, J.A. and Bowen-Pope, D.F. (1992): Regulation and role of PDGF receptor alpha-subunit expression during embryogenesis. Development, 115: 123–131
Scherson, T., Serbedzija, G., Fraser, S. and Bronner-Fraser, M. (1993): Regulative capacity of the cranial neural tube to form neural crest. Development, 118: 1049–1062
Schier, A.F., Neuhauss, S.C., Helde, K.A., Talbot, W.S. and Driever, W. (1997): The one-eyed pinhead gene functions in mesoderm and endoderm formation in zebrafish and interacts with no tail. Development, 124: 327–342
Schneider, R.A. (1999): Neural crest can form cartilages normally derived from mesoderm during development of the avian head skeleton. Dev Biol, 208: 441–455
Schneider, V.A. and Mercola, M. (1999): Spatially distinct head and heart inducers within the Xenopus organizer region. Curr Biol, 9: 800–809
Schneider, R.A., Hu, D., Rubenstein, J.L., Maden, M. and Helms, J.A. (2001): Local retinoid signaling coordinates forebrain and facial morphogenesis by maintaining FGF8 and SHH. Development, 128: 2755–2767
Schoenwolf, G.C., Folsom, D. and Moe, A. (1988): A reexamination of the role of microfilaments in neurulation in the chick embryo. Anat Rec, 220: 87–102
Schoenwolf, G.C. and Smith, J.L. (1990): Mechanisms of neurulation: traditional viewpoint and recent advances. Development, 109: 243–270
Schorle, H., Meier, P., Buchert, M., Jaenisch, R. and Mitchell, P.J. (1996): Transcription factor AP-2 essential for cranial closure and craniofacial development. Nature, 381: 235–238
Schuffenhauer, S., Lichtner, P., Peykar-Derakhshandeh, P., Murken, J., Haas, O.A., Back, E., Wolff, G., Zabel, B., Barisic, I., Rauch, A., Borochowitz, Z., Dallapiccola, B., Ross, M. and Meitinger, T. (1998): Deletion mapping on chromosome 10p and definition of a critical region for the second DiGeorge syndrome locus (DGS2). Eur J Hum Genet, 6: 213–225
Schlüter, G. (1973): Ultrastructural observations on cell necrosis during formation of the neural tube in mouse embryos. Z Anat Entwicklungsgesch, 141:251–64
Schutte, B.C. and Murray, J.C. (1999): The many faces and factors of orofacial clefts. Hum Mol Genet, 8: 1853–1859
Schweiger, S., Foerster, J., Lehmann, T., Suckow, V., Muller, Y.A., Walter, G., Davies, T., Porter, H., van Bokhoven, H., Lunt, P.W., Traub, P. and Ropers, H.H. (1999): The Opitz syndrome gene product, MID1, associates with microtubules. Proc Natl Acad Sci U S A, 96: 2794–2799

Sciote, J.J., Rowlerson, A.M., Hopper, C. and Hunt, N.P. (1994): Fibre type classification and myosin isoforms in the human masseter muscle. J Neurol Sci, 126: 15–24
Sciote, J.J., Rowlerson, A.M. and Carlson, D.S. (1995): Myosin expression in the jaw-closing muscles of the domestic cat and American opossum. Arch Oral Biol, 40: 405–413
Sechrist, J., Serbedzija, G.N., Scherson, T., Fraser, S.E. and Bronner-Fraser, M. (1993): Segmental migration of the hindbrain neural crest does not arise from its segmental generation. Development, 118: 691–703
Sechrist, J., Scherson, T. and Bronner-Fraser, M. (1994): Rhombomere rotation reveals that multiple mechanisms contribute to the segmental pattern of hindbrain neural crest migration. Development, 120: 1777–1790
Sefton, M., Sanchez, S. and Nieto, M.A. (1998): Conserved and divergent roles for members of the Snail family of transcription factors in the chick and mouse embryo. Development, 125: 3111–3121
Selleck, M.A. and Bronner-Fraser, M. (1995): Origins of the avian neural crest: the role of neural plate-epidermal interactions. Development, 121: 525–538
Selleck, M.A. and Bronner-Fraser, M. (1996): The genesis of avian neural crest cells: a classic embryonic induction. Proc Natl Acad Sci U S A, 93: 9352–9357
Semina, E.V., Reiter, R., Leysens, N.J., Alward, W.L., Small, K.W., Datson, N.A., Siegel-Bartelt, J., Bierke-Nelson, D., Bitoun, P., Zabel, B.U., Carey, J.C. and Murray, J.C. (1996): Cloning and characterization of a novel bicoid-related homeobox transcription factor gene, RIEG, involved in Rieger syndrome. Nat Genet, 14: 392–399
Serbedzija, G.N., Bronner-Fraser, M. and Fraser, S.E. (1992): Vital dye analysis of cranial neural crest cell migration in the mouse embryo. Development, 116: 297–307
Sertie, A.L., Sossi, V., Camargo, A.A., Zatz, M., Brahe, C. and Passos-Bueno, M.R. (2000): Collagen XVIII, containing an endogenous inhibitor of angiogenesis and tumor growth, plays a critical role in the maintenance of retinal structure and in neural tube closure (Knobloch syndrome). Hum Mol Genet, 9: 2051–2058
Sham, M.H., Vesque, C., Nonchev, S., Marshall, H., Frain, M., Gupta, R.D., Whiting, J., Wilkinson, D., Charnay, P. and Krumlauf, R. (1993): The zinc finger gene Krox20 regulates HoxB2 (Hox2.8) during hindbrain segmentation. Cell, 72: 183–196
Shankar, K.R., Chuong, C.M., Jaskoll, T. and Melnick, M. (1994): Effect of in ovo retinoic acid exposure on forebrain neural crest: in vitro analysis reveals up-regulation of N-CAM and loss of mesenchymal phenotype. Dev Dyn, 200: 89–102
Shelton, G.D., Cardinet, G.H., 3rd and Bandman, E. (1988): Expression of fiber type specific proteins during ontogeny of canine temporalis muscle. Muscle Nerve, 11: 124–132
Shigetani, Y., Nobusada, Y. and Kuratani, S. (2000): Ectodermally derived FGF8 defines the maxillomandibular region in the early chick embryo: epithelial-mesenchymal interactions in the specification of the craniofacial ectomesenchyme. Dev Biol, 228: 73–85
Shimamura, K., Hartigan, D.J., Martinez, S., Puelles, L. and Rubenstein, J.L. (1995): Longitudinal organization of the anterior neural plate and neural tube. Development, 121: 3923–3933
Shuler, C.F. (1995): Programmed cell death and cell transformation in craniofacial development. Crit Rev Oral Biol Med, 6: 202–217
Sidow, A., Bulotsky, M.S., Kerrebrock, A.W., Bronson, R.T., Daly, M.J., Reeve, M.P., Hawkins, T.L., Birren, B.W., Jaenisch, R. and Lander, E.S. (1997): Serrate2 is disrupted in the mouse limb-development mutant syndactylism. Nature, 389: 722–725
Smith, D.E., Franco del Amo, F., Gridley, T. (1992): Isolation of Sna, a mouse gene homologous to the Drosophila genes snail and escargot: its expression pattern suggests multiple roles during postimplantation development. Development, 116:1033–9
Smith, M.M. and Hall, B.K. (1993) A developmental model for evolution of the vertebrate exoskeleton and teeth. In Hecht, M.K. (ed.), *Evolutionary Biology*,, Vol. 27, pp. 387–448, New York: Plenum Press
Smith, A., Robinson, V., Patel, K. and Wilkinson, D.G. (1997): The EphA4 and EphB1 receptor tyrosine kinases and ephrin-B2 ligand regulate targeted migration of branchial neural crest cells. Curr Biol, 7: 561–570
Smith, J.L. and Schoenwolf, G.C. (1997): Neurulation: coming to closure. Trends Neurosci, 20: 510–517
Smith, K.K. and Schneider, R.A. (1998): Have gene knockouts caused evolutionary reversals in the mammalian first arch? Bioessays, 20: 245–255

Smith, A. and Graham, A. (2001): Restricting Bmp-4 mediated apoptosis in hindbrain neural crest. Dev Dyn, 220: 276–283

Smitherman, M., Lee, K., Swanger, J., Kapur, R. and Clurman, B.E. (2000): Characterization and targeted disruption of murine Nup50, a p27(Kip1)- interacting component of the nuclear pore complex. Mol Cell Biol, 20: 5631–5642

Sohal, G.S., Ali, A.A. and Ali, M.M. (1998): Ventral neural tube cells differentiate into craniofacial skeletal muscles. Biochem Biophys Res Commun, 252: 675–678

Sohal, G.S., Ali, M.M., Ali, A.A. and Dai, D. (1999a): Ventrally emigrating neural tube cells contribute to the formation of Meckel's and quadrate cartilage. Dev Dyn, 216: 37–44

Sohal, G.S., Ali, M.M., Ali, A.A. and Dai, D. (1999b): Ventrally emigrating neural tube cells differentiate into heart muscle. Biochem Biophys Res Commun, 254: 601–604

Sohal, G.S., Bockman, D.E., Ali, M.M. and Tsai, N.T. (1996): DiI labeling and homeobox gene islet-1 expression reveal the contribution of ventral neural tube cells to the formation of the avian trigeminal ganglion. Int J Dev Neurosci, 14: 419–427

Solloway, M.J. and Robertson, E.J. (1999): Early embryonic lethality in Bmp5/Bmp7 double mutant mice suggests functional redundancy within the 60 A subgroup. Development, 126: 1753–1768

Soriano, P. (1997): The PDGF alpha receptor is required for neural crest cell development and for normal patterning of the somites. Development, 124: 2691–2700

Soussi-Yanicostas, N., Barbet, J.P., Laurent-Winter, C., Barton, P., Butler-Browne, G.S. (1990): Transition of myosin isozymes during development of human masseter muscle. Persistence of developmental isoforms during postnatal stage. Development. 108:239–49

Sozen, M.A., Suzuki, K., Tolarova, M.M., Bustos, T., Fernandez Iglesias, J.E., Spritz, R.A. (2001): Mutation of PVRL1 is associated with sporadic, non-syndromic cleft lip/palate in northern Venezuela. Nat Genet, 29:141–2

Spector, M.S., Raff, A., DeSilva, H., Lee, K. and Osley, M.A. (1997): Hir1p and Hir2p function as transcriptional corepressors to regulate histone gene transcription in the Saccharomyces cerevisiae cell cycle. Mol Cell Biol, 17: 545–552

Spokony, R.F., Aoki, Y., Saint-German, N., Magner-Fink, E. and Saint-Jeannet, J-P. (2002): The transcription factor Sox9 is required for cranial neural crest development in Xenopus. Development, 129: 421–432.

Srivastava, D., Thomas, T., Lin, Q., Kirby, M.L., Brown, D. and Olson, E.N. (1997): Regulation of cardiac mesodermal and neural crest development by the bHLH transcription factor, dHAND. Nat Genet, 16: 154–160

Stockton, D.W., Das, P., Goldenberg, M., D'Souza, R.N. and Patel, P.I. (2000): Mutation of PAX9 is associated with oligodontia. Nat Genet, 24: 18–19.

Stoetzel, C., Weber, B., Bourgeois, P., Bolcato-Bellemin, A.L. and Perrin-Schmitt, F. (1995): Dorsoventral and rostro-caudal sequential expression of M-twist in the postimplantation murine embryo. Mech Dev, 51: 251–263

Stottmann, R.W., Anderson, R.M. and Klingensmith, J. (2001): The BMP Antagonists Chordin and Noggin Have Essential but Redundant Roles in Mouse Mandibular Outgrowth. Dev Biol, 240: 457–473

Strähle, U., Fischer, N. and Blader, P. (1997): Expression and regulation of a netrin homologue in the zebrafish embryo. Mech Dev, 62: 147–160

Streit, A. and Stern, C.D. (1999): Establishment and maintenance of the border of the neural plate in the chick: involvement of FGF and BMP activity. Mech Dev, 82: 51–66

Stumpo, D.J., Bock, C.B., Tuttle, J.S. and Blackshear, P.J. (1995): MARCKS deficiency in mice leads to abnormal brain development and perinatal death. Proc Natl Acad Sci U S A, 92: 944–948

Sucov, H.M., Izpisua-Belmonte, J.C., Ganan, Y. and Evans, R.M. (1995): Mouse embryos lacking RXR alpha are resistant to retinoic-acid-induced limb defects. Development, 121: 3997–4003

Sulik, K., Dehart, D.B., Iangaki, T., Carson, J.L., Vrablic, T., Gesteland, K. and Schoenwolf, G.C. (1994): Morphogenesis of the murine node and notochordal plate. Dev Dyn, 201: 260–278

Sulik, K.K., Dehart, D.B., Rogers, J.M. and Chernoff, N. (1995): Teratogenicity of low doses of all-trans retinoic acid in presomite mouse embryos. Teratology, 51: 398–403

Sullivan, R., Huang, G.Y., Meyer, R.A., Wessels, A., Linask, K.K. and Lo, C.W. (1998): Heart malformations in transgenic mice exhibiting dominant negative inhibition of gap junctional communication in neural crest cells. Dev Biol, 204: 224–234

Summerbell, D., Ashby, P.R., Coutelle, O., Cox, D., Yee, S. and Rigby, P.W. (2000): The expression of Myf5 in the developing mouse embryo is controlled by discrete and dispersed enhancers specific for particular populations of skeletal muscle precursors. Development, 127: 3745–3757

Sun, D., Vanderburg, C.R., Odierna, G.S. and Hay, E.D. (1998): TGFbeta3 promotes transformation of chicken palate medial edge epithelium to mesenchyme in vitro. Development, 125: 95–105

Sun, T., Jayatilake, D., Afink, G.B., Ataliotis, P., Nister, M., Richardson, W.D. and Smith, H.K. (2000): A human YAC transgene rescues craniofacial and neural tube development in PDGFRalpha knockout mice and uncovers a role for PDGFRalpha in prenatal lung growth. Development, 127: 4519–4529

Suzuki, K., Hu, D., Bustos, T., Zlotogora, J., Richieri-Costa, A., Helms, J.A. and Spritz, R.A. (2000): Mutations of PVRL1, encoding a cell-cell adhesion molecule/herpesvirus receptor, in cleft lip/palate-ectodermal dysplasia. Nat Genet, 25: 427–430

Swindell, E.C., Thaller, C., Sockanathan, S., Petkovich, M., Jessell, T.M. and Eichele, G. (1999): Complementary domains of retinoic acid production and degradation in the early chick embryo. Dev Biol, 216: 282–296

Szeto, D.P., Rodriguez-Esteban, C., Ryan, A.K., O'Connell, S.M., Liu, F., Kioussi, C., Gleiberman, A.S., Izpisua-Belmonte, J.C. and Rosenfeld, M.G. (1999): Role of the Bicoid-related homeodomain factor Pitx1 in specifying hindlimb morphogenesis and pituitary development. Genes Dev, 13: 484–494

Tada, M. and Smith, J.C. (2000): Xwnt11 is a target of Xenopus Brachyury: regulation of gastrulation movements via Dishevelled, but not through the canonical Wnt pathway. Development, 127: 2227–2238

Takahashi, K., Nuckolls, G.H., Tanaka, O., Semba, I., Takahashi, I., Dashner, R., Shum, L. and Slavkin, H.C. (1998): Adenovirus-mediated ectopic expression of Msx2 in even-numbered rhombomeres induces apoptotic elimination of cranial neural crest cells in ovo. Development, 125: 1627–1635

Takahashi, K., Nuckolls, G.H., Takahashi, K., Nonaka, K., Nagata, M., Ikura, T., Slavkin, H.C. and Shum, L. (2001): Msx2 is a repressor of chondrogenic differentiation in migratory cranial neural crest cells. Dev Dyn, 222: 252–262

Takagi, T., Moribe, H., Kondoh, H. and Higashi, Y. (1998): DeltaEF1, a zinc finger and homeodomain transcription factor, is required for skeleton patterning in multiple lineages. Development, 125: 21–31

Takeuchi, T., Yamazaki, Y., Katoh-Fukui, Y., Tsuchiya, R., Kondo, S., Motoyama, J. and Higashinakagawa, T. (1995): Gene trap capture of a novel mouse gene, jumonji, required for neural tube formation. Genes Dev, 9: 1211–1222

Takeuchi, T., Kojima, M., Nakajima, K. and Kondo, S. (1999): Jumonji gene is essential for the neurulation and cardiac development of mouse embryos with a C3H/He background. Mech Dev, 86: 29–38

Takihara, Y., Tomotsune, D., Shirai, M., Katoh-Fukui, Y., Nishii, K., Motaleb, M.A., Nomura, M., Tsuchiya, R., Fujita, Y., Shibata, Y., Higashinakagawa, T. and Shimada, K. (1997): Targeted disruption of the mouse homologue of the Drosophila polyhomeotic gene leads to altered antero-posterior patterning and neural crest defects. Development, 124: 3673–3682

Tam, P.P., Meier, S. and Jacobson, A.G. (1982): Differentiation of the metameric pattern in the embryonic axis of the mouse. II. Somitomeric organization of the presomitic mesoderm. Differentiation, 21: 109–122

Tam, J.P., Liu, W., Zhang, J.W., Galantino, M., Bertolero, F., Cristiani, C., Vaghi, F. and de Castiglione, R. (1994): Alanine scan of endothelin: importance of aromatic residues. Peptides, 15: 703–708

Tam, P.P. and Trainor, P.A. (1994): Specification and segmentation of the paraxial mesoderm. Anat Embryol (Berl), 189: 275–305

Tan, S.S. and Morriss-Kay, G. (1985): The development and distribution of the cranial neural crest in the rat embryo. Cell Tissue Res, 240: 403–416

Tan, C., Deardorff, M.A., Saint-Jeannet, J.P., Yang, J., Arzoumanian, A. and Klein, P.S. (2001): Kermit, a frizzled interacting protein, regulates frizzled 3 signaling in neural crest development. Development, 128: 3665–3674

Taneja, R., Thisse, B., Rijli, F.M., Thisse, C., Bouillet, P., Dolle, P. and Chambon, P. (1996): The expression pattern of the mouse receptor tyrosine kinase gene MDK1 is conserved through evolution and requires Hoxa-2 for rhombomere- specific expression in mouse embryos. Dev Biol, 177: 397–412

Tassabehji, M., Read, A.P., Newton, V.E., Patton, M., Gruss, P., Harris, R. and Strachan, T. (1993): Mutations in the PAX3 gene causing Waardenburg syndrome type 1 and type 2. Nat Genet, 3: 26–30

Taya, Y., O'Kane, S. and Ferguson, M.W. (1999): Pathogenesis of cleft palate in TGF-beta3 knockout mice. Development, 126: 3869–3879

ten Berge, D., Brouwer, A., Korving, J., Martin, J.F. and Meijlink, F. (1998): Prx1 and Prx2 in skeletogenesis: roles in the craniofacial region, inner ear and limbs. Development, 125: 3831–3842

ten Berge, D., Brouwer, A., Korving, J., Reijnen, M.J., van Raaij, E.J., Verbeek, F., Gaffield, W. and Meijlink, F. (2001): Prx1 and Prx2 are upstream regulators of sonic hedgehog and control cell proliferation during mandibular arch morphogenesis. Development, 128: 2929–2938

Thesleff, I. and Sharpe, P. (1997): Signalling networks regulating dental development. Mech Dev, 67: 111–123

Thisse, C., Thisse, B., Halpern, M.E. and Postlethwait, J.H. (1994): Goosecoid expression in neurectoderm and mesendoderm is disrupted in zebrafish cyclops gastrulas. Dev Biol, 164: 420–429

Thisse, C. and Thisse, B. (1999): Antivin, a novel and divergent member of the TGFbeta superfamily, negatively regulates mesoderm induction. Development, 126: 229–240

Thomas, T., Kurihara, H., Yamagishi, H., Kurihara, Y., Yazaki, Y., Olson, E.N. and Srivastava, D. (1998): A signaling cascade involving endothelin-1, dHAND and msx1 regulates development of neural-crest-derived branchial arch mesenchyme. Development, 125: 3005–3014

Thomas, B.L., Liu, J.K., Rubenstein, J.L. and Sharpe, P.T. (2000): Independent regulation of Dlx2 expression in the epithelium and mesenchyme of the first branchial arch. Development, 127: 217–224

Tosney, K.W. (1982): The segregation and early migration of cranial neural crest cells in the avian embryo. Dev Biol, 89: 13–24

Trainor, P.A., Tan, S.S. and Tam, P.P. (1994): Cranial paraxial mesoderm: regionalisation of cell fate and impact on craniofacial development in mouse embryos. Development, 120: 2397–2408

Trainor, P.A. and Tam, P.P. (1995): Cranial paraxial mesoderm and neural crest cells of the mouse embryo: co-distribution in the craniofacial mesenchyme but distinct segregation in branchial arches. Development, 121: 2569–2582

Trainor, P. and Krumlauf, R. (2000): Plasticity in mouse neural crest cells reveals a new patterning role for cranial mesoderm. Nat Cell Biol, 2: 96–102

Trainor, P.A., Ariza-McNaughton, L. and Krumlauf, R. (2002a): Role of the isthmus and FGFs in resolving the paradox of neural crest plasticity and prepatterning. Science, 295: 1288–1291

Trainor, P.A., Sobieszczuk, D., Wilkinson, D. and Krumlauf, R. (2002b): Signalling between the hindbrain and paraxial tissues dictates neural crest migration pathways. Development, 129: 433–442.

Tribioli, C. and Lufkin, T. (1999): The murine Bapx1 homeobox gene plays a critical role in embryonic development of the axial skeleton and spleen. Development, 126: 5699–5711

Trumpp, A., Depew, M.J., Rubenstein, J.L., Bishop, J.M. and Martin, G.R. (1999): Cre-mediated gene inactivation demonstrates that FGF8 is required for cell survival and patterning of the first branchial arch. Genes Dev, 13: 3136–3148

Tucker, A.S., Al Khamis, A. and Sharpe, P.T. (1998a): Interactions between Bmp-4 and Msx-1 act to restrict gene expression to odontogenic mesenchyme. Dev Dyn, 212: 533–539

Tucker, A.S., Matthews, K.L. and Sharpe, P.T. (1998b): Transformation of tooth type induced by inhibition of BMP signaling. Science, 282: 1136–1138

Tucker, A.S., Yamada, G., Grigoriou, M., Pachnis, V. and Sharpe, P.T. (1999): Fgf-8 determines rostral-caudal polarity in the first branchial arch. Development, 126: 51–61

Uchiyama, K., Ishikawa, A. and Hanaoka, K. (2000): Expression of lbx1 involved in the hypaxial musculature formation of the mouse embryo. J Exp Zool, 286: 270–279

Van Allen, M.I., Kalousek, D.K., Chernoff, G.F., Juriloff, D., Harris, M., McGillivray, B.C., Yong, S.L., Langlois, S., MacLeod, P.M., Chitayat, D. and et al. (1993): Evidence for multi-site closure of the neural tube in humans. Am J Med Genet, 47: 723–743

van Bokhoven, H., Hamel, B.C., Bamshad, M., Sangiorgi, E., Gurrieri, F., Duijf, P.H., Vanmolkot, K.R., van Beusekom, E., van Beersum, S.E., Celli, J., Merkx, G.F., Tenconi, R., Fryns, J.P., Verloes, A., Newbury-Ecob, R.A., Raas-Rotschild, A., Majewski, F., Beemer, F.A., Janecke, A., Chitayat, D., Crisponi, G., Kayserili, H., Yates, J.R., Neri, G. and Brunner, H.G. (2001): p63 Gene mutations in

EEC syndrome, limb-mammary syndrome, and isolated split hand-split foot malformation suggest a genotype-phenotype correlation. Am J Hum Genet, 69: 481–492
van den Boogaard, M.J., Dorland, M., Beemer, F.A. and van Amstel, H.K. (2000): MSX1 mutation is associated with orofacial clefting and tooth agenesis in humans. Nat Genet, 24: 342–343
van der Put, N.M., van Straaten, H.W., Trijbels, F.J. and Blom, H.J. (2001): Folate, homocysteine and neural tube defects: an overview. Exp Biol Med, 226: 243–270
Varga, Z.M., Wegner, J. and Westerfield, M. (1999): Anterior movement of ventral diencephalic precursors separates the primordial eye field in the neural plate and requires cyclops. Development, 126: 5533–5546
Vastardis, H., Karimbux, N., Guthua, S.W., Seidman, J.G. and Seidman, C.E. (1996): A human MSX1 homeodomain missense mutation causes selective tooth agenesis. Nat Genet, 13: 417–421
Veitch, E., Begbie, J., Schilling, T.F., Smith, M.M. and Graham, A. (1999): Pharyngeal arch patterning in the absence of neural crest. Curr Biol, 9: 1481–1484.1
Vortkamp, A., Gessler, M. and Grzeschik, K.H. (1991): GLI3 zinc-finger gene interrupted by translocations in Greig syndrome families. Nature, 352: 539–540
Wachtler, F. and Jacob, M. (1986): Origin and development of the cranial skeletal muscles. Bibl Anat, 29: 24–46
Wadey, R., McKie, J., Papapetrou, C., Sutherland, H., Lohman, F., Osinga, J., Frohn, I., Hofstra, R., Meijers, C., Amati, F., Conti, E., Pizzuti, A., Dallapiccola, B., Novelli, G. and Scambler, P. (1999): Mutations of UFD1L are not responsible for the majority of cases of DiGeorge Syndrome/velocardiofacial syndrome without deletions within chromosome 22q11. Am J Hum Genet, 65: 247–249
Wagner, G. (1949): Die bedeutung der neuralleiste für die kopfgestaltung der amphibien larven. Rev Suisse Zool, 56: 519–620
Wahl, C.M., Noden, D.M. and Baker, R. (1994): Developmental relations between sixth nerve motor neurons and their targets in the chick embryo. Dev Dyn, 201: 191–202
Waldo, K.L., Kumiski, D. and Kirby, M.L. (1996): Cardiac neural crest is essential for the persistence rather than the formation of an arch artery. Dev Dyn, 205: 281–292
Waldrip, W.R., Bikoff, E.K., Hoodless, P.A., Wrana, J.L. and Robertson, E.J. (1998): Smad2 signaling in extraembryonic tissues determines anterior-posterior polarity of the early mouse embryo. Cell, 92: 797–808
Wall, N.A. and Hogan, B.L. (1995): Expression of bone morphogenetic protein-4 (BMP-4), bone morphogenetic protein-7 (BMP-7), fibroblast growth factor-8 (FGF-8) and sonic hedgehog (SHH) during branchial arch development in the chick. Mech Dev, 53: 383–392
Wallis, D.E., Roessler, E., Hehr, U., Nanni, L., Wiltshire, T., Richieri-Costa, A., Gillessen-Kaesbach, G., Zackai, E.H., Rommens, J. and Muenke, M. (1999): Mutations in the homeodomain of the human SIX3 gene cause holoprosencephaly. Nat Genet, 22: 196–198
Walther, C. and Gruss, P. (1991): Pax-6, a murine paired box gene, is expressed in the developing CNS. Development, 113: 1435–1449
Wang, H.U., Chen, Z.F. and Anderson, D.J. (1998): Molecular distinction and angiogenic interaction between embryonic arteries and veins revealed by ephrin-B2 and its receptor Eph-B4. Cell, 93: 741–753
Wassif, C.A., Zhu, P., Kratz, L., Krakowiak, P.A., Battaile, K.P., Weight, F.F., Grinberg, A., Steiner, R.D., Nwokoro, N.A., Kelley, R.I., Stewart, R.R. and Porter, F.D. (2001): Biochemical, phenotypic and neurophysiological characterization of a genetic mouse model of RSH/Smith–Lemli–Opitz syndrome. Hum Mol Genet, 10: 555–564
Watanabe, T., Goulding, E.H. and Pratt, R.M. (1988): Alterations in craniofacial growth induced by isotretinoin (13-cis- retinoic acid) in mouse whole embryo and primary mesenchymal cell culture. J Craniofac Genet Dev Biol, 8: 21–33
Watanabe, H. and Yamada, Y. (1999): Mice lacking link protein develop dwarfism and craniofacial abnormalities. Nat Genet, 21: 225–229
Weaver, M., Dunn, N.R. and Hogan, B.L. (2000): Bmp4 and Fgf10 play opposing roles during lung bud morphogenesis. Development, 127: 2695–2704
Wedden, S.E. and Tickle, C. (1986): Facial morphogenesis and pattern formation. Prog Clin Biol Res, 335–337
Wedden, S.E. (1987): Epithelial-mesenchymal interactions in the development of chick facial primordia and the target of retinoid action. Development, 99: 341–351

Wee, E.L., Babiarz, B.S., Zimmerman, S. and Zimmerman, E.F. (1979): Palate morphogenesis. IV. Effects of serotonin and its antagonists on rotation in embryo culture. J Embryol Exp Morphol, 53: 75–90
Wee, E.L. and Zimmerman, E.F. (1983): Involvement of GABA in palate morphogenesis and its relation to diazepam teratogenesis in two mouse strains. Teratology, 28: 15–22
Weil, M., Jacobson, M.D. and Raff, M.C. (1997): Is programmed cell death required for neural tube closure? Curr Biol, 7: 281–284
Weinstein, M., Yang, X., Li, C., Xu, X., Gotay, J. and Deng, C.X. (1998): Failure of egg cylinder elongation and mesoderm induction in mouse embryos lacking the tumor suppressor smad2. Proc Natl Acad Sci U S A, 95: 9378–9383
Wendling, O., Chambon, P. and Mark, M. (1999): Retinoid X receptors are essential for early mouse development and placentogenesis. Proc Natl Acad Sci U S A, 96: 547–551
Wigmore, P.M. and Evans, D.J.R. (2002): Molecular and cellular mechanisms involved in the generation of fiber diversity during myogenesis. Int. Rev. Cytol. 216:175–232
Wild, A., Kalff-Suske, M., Vortkamp, A., Bornholdt, D., Konig, R. and Grzeschik, K.H. (1997): Point mutations in human GLI3 cause Greig syndrome. Hum Mol Genet, 6: 1979–1984
Wilkie, A.O., Tang, Z., Elanko, N., Walsh, S., Twigg, S.R., Hurst, J.A., Wall, S.A., Chrzanowska, K.H. and Maxson, R.E., Jr. (2000): Functional haploinsufficiency of the human homeobox gene MSX2 causes defects in skull ossification. Nat Genet, 24: 387–390
Wilkie, A.O., Oldridge, M., Tang, Z. and Maxson, R.E., Jr. (2001): Craniosynostosis and related limb anomalies. Novartis Found Symp, 232: 122–133
Willnow, T.E., Hilpert, J., Armstrong, S.A., Rohlmann, A., Hammer, R.E., Burns, D.K. and Herz, J. (1996): Defective forebrain development in mice lacking gp330/megalin. Proc Natl Acad Sci U S A, 93: 8460–8464
Wilming, L.G., Snoeren, C.A., van Rijswijk, A., Grosveld, F. and Meijers, C. (1997): The murine homologue of HIRA, a DiGeorge syndrome candidate gene, is expressed in embryonic structures affected in human CATCH22 patients. Hum Mol Genet, 6: 247–258
Wilson, J.B., Ferguson, M.W., Jenkins, N.A., Lock, L.F., Copeland, N.G. and Levine, A.J. (1993): Transgenic mouse model of X-linked cleft palate. Cell Growth Differ, 4: 67–76
Winnier, G., Blessing, M., Labosky, P.A. and Hogan, B.L. (1995): Bone morphogenetic protein-4 is required for mesoderm formation and patterning in the mouse. Genes Dev, 9: 2105–2116
Winokur, S.T. and Shiang, R. (1998): The Treacher Collins syndrome (TCOF1) gene product, treacle, is targeted to the nucleolus by signals in its C-terminus. Hum Mol Genet, 7: 1947–1952
Wise, C.A., Chiang, L.C., Paznekas, W.A., Sharma, M., Musy, M.M., Ashley, J.A., Lovett, M. and Jabs, E.W. (1997): TCOF1 gene encodes a putative nucleolar phosphoprotein that exhibits mutations in Treacher Collins Syndrome throughout its coding region. Proc Natl Acad Sci U S A, 94: 3110–3115
Wolf, C., Thisse, C., Stoetzel, C., Thisse, B., Gerlinger, P. and Perrin-Schmitt, F. (1991): The M-twist gene of Mus is expressed in subsets of mesodermal cells and is closely related to the Xenopus X-twi and the Drosophila twist genes. Dev Biol, 143: 363–373
Woo, M., Hakem, R., Soengas, M.S., Duncan, G.S., Shahinian, A., Kagi, D., Hakem, A., McCurrach, M., Khoo, W., Kaufman, S.A., Senaldi, G., Howard, T., Lowe, S.W., Mak, T.W. (1998): Essential contribution of caspase 3/CPP32 to apoptosis and its associated nuclear changes. Genes Dev, 12:806–19
Wotton, D., Lo, R.S., Lee, S. and Massague, J. (1999a): A Smad transcriptional corepressor. Cell, 97: 29–39
Wotton, D., Lo, R.S., Swaby, L.A. and Massague, J. (1999b): Multiple modes of repression by the Smad transcriptional corepressor TGIF. J Biol Chem, 274: 37105–37110
Wragg, L.E., Smith, J.A. and Borden, C.S. (1972): Myoneural maturation and function of the foetal rat tongue at the time of secondary palate closure. Arch Oral Biol, 17: 673–682
Wu, M., Chen, D.F., Sasaoka, T. and Tonegawa, S. (1996): Neural tube defects and abnormal brain development in F52-deficient mice. Proc Natl Acad Sci U S A, 93: 2110–2115
Wu, Y.Q., Badano, J.L., McCaskill, C., Vogel, H., Potocki, L. and Shaffer, L.G. (2000): Haploinsufficiency of ALX4 as a potential cause of parietal foramina in the 11p11.2 contiguous gene-deletion syndrome. Am J Hum Genet, 67: 1327–1332
Wunsch, A.M. and Haas, A.L. (1995): Ubiquitin-protein conjugates selectively distribute during early chicken embryogenesis. Dev Dyn, 204: 118–132

Wuyts, W., Cleiren, E., Homfray, T., Rasore-Quartino, A., Vanhoenacker, F. and Van Hul, W. (2000): The ALX4 homeobox gene is mutated in patients with ossification defects of the skull (foramina parietalia permagna, OMIM 168500). J Med Genet, 37: 916–920

Xu, X., Li, W.E., Huang, G.Y., Meyer, R., Chen, T., Luo, Y., Thomas, M.P., Radice, G.L. and Lo, C.W. (2001): Modulation of mouse neural crest cell motility by N-cadherin and connexin 43 gap junctions. J Cell Biol, 154: 217–230

Yamada, G., Mansouri, A., Torres, M., Stuart, E.T., Blum, M., Schultz, M., De Robertis, E.M. and Gruss, P. (1995): Targeted mutation of the murine goosecoid gene results in craniofacial defects and neonatal death. Development, 121: 2917–2922

Yamada, G., Ueno, K., Nakamura, S., Hanamure, Y., Yasui, K., Uemura, M., Eizuru, Y., Mansouri, A., Blum, M. and Sugimura, K. (1997): Nasal and pharyngeal abnormalities caused by the mouse goosecoid gene mutation. Biochem Biophys Res Commun, 233: 161–165

Yamagishi, H., Garg, V., Matsuoka, R., Thomas, T. and Srivastava, D. (1999): A molecular pathway revealing a genetic basis for human cardiac and craniofacial defects. Science, 283: 1158–1161

Yamaguchi, T.P., Bradley, A., McMahon, A.P. and Jones, S. (1999): A Wnt5a pathway underlies outgrowth of multiple structures in the vertebrate embryo. Development, 126: 1211–1223

Yamane, A., Mayo, M.L., Bringas, P. Jr., Chen, L., Huynh, M., Thai, K., Shum, L., Slavkin, H.C. (1997): TGF-alpha, EGF, and their cognate EGF receptor are co-expressed with desmin during embryonic, fetal, and neonatal myogenesis in mouse tongue development. Dev Dyn, 209:353-66

Yamane, A., Bringas, P. Jr, Mayo, M.L., Amano, O., Takahashi, K., Vo, H., Shum, L., Slavkin, H.C. (1998): Transforming growth factor alpha up-regulates desmin expression during embryonic mouse tongue myogenesis. Dev Dyn, 213:71-81

Yamamoto, T.S., Takagi, C. and Ueno, N. (2000): Requirement of Xmsx-1 in the BMP-triggered ventralization of Xenopus embryos. Mech Dev, 91: 131–141

Yamauchi, Y., Abe, K., Mantani, A., Hitoshi, Y., Suzuki, M., Osuzu, F., Kuratani, S. and Yamamura, K. (1999): A novel transgenic technique that allows specific marking of the neural crest cell lineage in mice. Dev Biol, 212: 191–203

Yanagisawa, H., Hammer, R.E., Richardson, J.A., Williams, S.C., Clouthier, D.E. and Yanagisawa, M. (1998a): Role of Endothelin-1/Endothelin-A receptor-mediated signaling pathway in the aortic arch patterning in mice. J Clin Invest, 102: 22–33

Yanagisawa, H., Yanagisawa, M., Kapur, R.P., Richardson, J.A., Williams, S.C., Clouthier, D.E., de Wit, D., Emoto, N. and Hammer, R.E. (1998b): Dual genetic pathways of endothelin-mediated intercellular signaling revealed by targeted disruption of endothelin converting enzyme-1 gene. Development, 125: 825–836

Yang, A., Schweitzer, R., Sun, D., Kaghad, M., Walker, N., Bronson, R.T., Tabin, C., Sharpe, A., Caput, D., Crum, C. and McKeon, F. (1999): p63 is essential for regenerative proliferation in limb, craniofacial and epithelial development. Nature, 398: 714–718

Yao, T.P., Oh, S.P., Fuchs, M., Zhou, N.D., Ch'ng, L.E., Newsome, D., Bronson, R.T., Li, E., Livingston, D.M. and Eckner, R. (1998): Gene dosage-dependent embryonic development and proliferation defects in mice lacking the transcriptional integrator p300. Cell, 93: 361-372

Ybot-Gonzalez, P. and Copp, A.J. (1999): Bending of the neural plate during mouse spinal neurulation is independent of actin microfilaments. Dev Dyn, 215: 273–283

Yoshida, H., Kong, Y.Y., Yoshida, R., Elia, A.J., Hakem, A., Hakem, R., Penninger, J.M. and Mak, T.W. (1998): Apafl is required for mitochondrial pathways of apoptosis and brain development. Cell, 94: 739–750

Yoshioka, H., Meno, C., Koshiba, K., Sugihara, M., Itoh, H., Ishimaru, Y., Inoue, T., Ohuchi, H., Semina, E.V., Murray, J.C., Hamada, H. and Noji, S. (1998): Pitx2, a bicoid-type homeobox gene, is involved in a lefty-signaling pathway in determination of left-right asymmetry. Cell, 94: 299–305

Yuan, Z.R., Kohsaka, T., Ikegaya, T., Suzuki, T., Okano, S., Abe, J., Kobayashi, N. and Yamada, M. (1998): Mutational analysis of the Jagged 1 gene in Alagille syndrome families. Hum Mol Genet, 7: 1363–1369

Zhang, J., Hagopian-Donaldson, S., Serbedzija, G., Elsemore, J., Plehn-Dujowich, D., McMahon, A.P., Flavell, R.A. and Williams, T. (1996): Neural tube, skeletal and body wall defects in mice lacking transcription factor AP-2. Nature, 381: 238–241

Zhang, P., Liegeois, N.J., Wong, C., Finegold, M., Hou, H., Thompson, J.C., Silverman, A., Harper, J.W., DePinho, R.A. and Elledge, S.J. (1997): Altered cell differentiation and proliferation in mice lacking p57KIP2 indicates a role in Beckwith-Wiedemann syndrome. Nature, 387: 151–158
Zhang, J., Talbot, W.S. and Schier, A.F. (1998): Positional cloning identifies zebrafish one-eyed pinhead as a permissive EGF-related ligand required during gastrulation. Cell, 92: 241–251
Zhao, Q., Behringer, R.R. and de Crombrugghe, B. (1996): Prenatal folic acid treatment suppresses acrania and meroanencephaly in mice mutant for the Cart1 homeobox gene. Nat Genet, 13: 275–283
Zhao, Y., Guo, Y.J., Tomac, A.C., Taylor, N.R., Grinberg, A., Lee, E.J., Huang, S. and Westphal, H. (1999): Isolated cleft palate in mice with a targeted mutation of the LIM homeobox gene lhx8. Proc Natl Acad Sci U S A, 96: 15002–15006
Zhao, F., Weismann, C.G., Satoda, M., Pierpont, M.E., Sweeney, E., Thompson, E.M. and Gelb, B.D. (2001): Novel TFAP2B mutations that cause Char syndrome provide a genotype- phenotype correlation. Am J Hum Genet, 69: 695–703
Zhong, W., Jiang, M.M., Schonemann, M.D., Meneses, J.J., Pedersen, R.A., Jan, L.Y. and Jan, Y.N. (2000): Mouse numb is an essential gene involved in cortical neurogenesis. Proc Natl Acad Sci U S A, 97: 6844–6849
Zhou, X., Sasaki, H., Lowe, L., Hogan, B.L. and Kuehn, M.R. (1993): Nodal is a novel TGF-beta-like gene expressed in the mouse node during gastrulation. Nature, 361: 543–547
Zhu, C.C., Yamada, G. and Blum, M. (1997): Correlation between loss of middle ear bones and altered goosecoid gene expression in the branchial region following retinoic acid treatment of mouse embryos in vivo. Biochem Biophys Res Commun, 235: 748–753
Zimmerman, E.F., Wee, E.L., Phillips, N. and Roberts, N. (1981): Presence of serotonin in the palate just prior to shelf elevation. J Embryol Exp Morphol, 64: 233–250
Zimmerman, E.F., Clark, R.L., Ganguli, S. and Venkatasubramanian, K. (1983): Serotonin regulation of palatal cell motility and metabolism. J Craniofac Genet Dev Biol, 3: 371–385
Zoltewicz, J.S. and Gerhart, J.C. (1997): The Spemann organizer of Xenopus is patterned along its anteroposterior axis at the earliest gastrula stage. Dev Biol, 192: 482–491
Zuniga, A., Haramis, A.P., McMahon, A.P. and Zeller, R. (1999): Signal relay by BMP antagonism controls the SHH/FGF4 feedback loop in vertebrate limb buds. Nature, 401: 598–602
Zutter, M.M., Ryan, E.E., Painter, A.D. (1997): Binding of phosphorylated Sp1 protein to tandem Sp1 binding sites regulates alpha2 integrin gene core promoter activity. Blood, 90:678–89

Subject Index

Production: Druckhaus Beltz, Hemsbach

9 783540 003632

GPSR Compliance

The European Union's (EU) General Product Safety Regulation (GPSR) is a set of rules that requires consumer products to be safe and our obligations to ensure this.

If you have any concerns about our products, you can contact us on ProductSafety@springernature.com

In case Publisher is established outside the EU, the EU authorized representative is:

Springer Nature Customer Service Center GmbH
Europaplatz 3
69115 Heidelberg, Germany

Batch number: 10704713

Printed by Printforce, the Netherlands